LONDON MATHEMATICAL SOCIETY LECTURE NOTE SERIES

Editor: PROFESSOR G. C. SHEPHARD, University of East Anglia

This series publishes the records of lectures and seminars on advanced topics in mathematics held at universities throughout the world. For the most part, these are at postgraduate level either presenting new material or describing older material in a new way. Exceptionally, topics at the undergraduate level may be published if the treatment is sufficiently original.

Prospective authors should contact the editor in the first instance.

Already published in this series

1. General cohomology theory and K-theory, PETER HILTON.
2. Numerical ranges of operators on normed spaces and of elements of normed algebras, F. F. BONSALL and J. DUNCAN.
3. Convex polytopes and the upper bound conjecture, P. McMULLEN and G. C. SHEPHARD.
4. Algebraic topology: A student's guide, J. F. ADAMS.
5. Commutative algebra, J. T. KNIGHT.
6. Finite groups of automorphisms, NORMAN BIGGS.
7. Introduction to combinatory logic, J. R. HINDLEY, B. LERCHER and J. P. SELDIN.
8. Integration and harmonic analysis on compact groups, R. E. EDWARDS.
9. Elliptic functions and elliptic curves, PATRICK DU VAL.
10. Numerical ranges II, F. F. BONSALL and J. DUNCAN.
11. New developments in topology, G. SEGAL (ed.).
12. Symposium on complex analysis Canterbury 1973, J. CLUNIE and W. K. HAYMAN (eds.).
13. Combinatorics, Proceedings of the British combinatorial conference 1973, T. P. McDONOUGH and V. C. MAVRON (eds.).
14. Analytic theory of abelian varieties, H. P. F. SWINNERTON-DYER.
15. An introduction to topological groups, P. J. HIGGINS.
16. Topics in finite groups, TERENCE M. GAGEN.
17. Differentiable germs and catastrophes, THEODOR BRÖCKER and L. LANDER.
18. A geometric approach to homology theory, S. BUONCRISTIANO, C. P. ROURKE and B. J. SANDERSON.
19. Graph theory, coding theory and block designs, P. J. CAMERON and J. H. VAN LINT.
20. Sheaf theory, B. R. TENNISON.
21. Automatic continuity, A. M. SINCLAIR.
22. Presentations of groups, D. L. JOHNSON.
23. Parallelisms of complete designs, PETER J. CAMERON.
24. The topology of Stiefel manifolds, I. M. JAMES.

London Mathematical Society Lecture Note Series. 25

Lie Groups and Compact Groups

JOHN F. PRICE

School of Mathematics
University of New South Wales
Kensington, Australia

CAMBRIDGE UNIVERSITY PRESS
CAMBRIDGE
LONDON NEW YORK MELBOURNE

Published by the Syndics of the Cambridge University Press
The Pitt Building, Trumpington Street, Cambridge CB2 1RP
Bentley House, 200 Euston Road, London NW1 2DB
32 East 57th Street, New York, N. Y. 10022, USA
296 Beaconsfield Parade, Middle Park, Melbourne 3206, Australia

First published 1977

Printed in Great Britain
at the University Press, Cambridge

Library of Congress Cataloguing in Publication Data

Price, John Frederick, 1943-
Lie groups and compact groups.

(London Mathematical Society Lecture note series; 25)
Bibliography: p.
Includes index.
1. Lie groups. 2. Compact groups. I. Title.
II. Series: London Mathematical Society. Lecture note series; 25.
QA387.P74 512'.55 76-14034

ISBN 0 521 21340 1

To Val
David, Matthew and Karen
and to Maharishi

Contents

Preface

The purpose of these notes is twofold: to provide a quick self-contained introduction to the general theory of Lie groups and to give the structure of compact connected groups and Lie groups in terms of certain distinguished 'simple' Lie groups. With regards to the first aim, the notes can be used to provide a general introduction to the fundamentals of Lie groups or as a bridge to more advanced texts. In either case, experience has shown that they are suitable for postgraduate students and, at least the earlier chapters, for senior undergraduates. Concerning the second aim, the existing treatments of the structure results referred to above seem to be all from a fairly advanced point of view (cf. Pontrjagin [1] and Weil [1]). It is hoped that the present, more modern treatment makes these powerful results more generally accessible, in particular to those only wishing to use them as a tool.

The theory of Lie groups lies at the junction of the theories of differentiable manifolds, topological groups and Lie algebras. In keeping with current trends, when dealing with manifolds (and hence with Lie groups) a coordinate-free notation is used, thus removing the necessity for tedious juggling of indices and, hopefully, adding to the clarity and intuitiveness of the theory. In the case of Lie groups, particular emphasis is placed upon results and techniques which educe the interplay between a Lie group and its Lie algebra.

During the past few years a number of important results have been obtained in harmonic analysis on compact groups and compact Lie groups by using the structure of these groups ... the overall orientation of the following notes is to give full details of several of these structure results. The main theorem for Lie groups is that if G is a compact connected Lie group, then G is topologically isomorphic to

$$(G_0 \times G_1 \times \dots \times G_m)/K,$$

where G_0 is the identity component of the centre of G, the G_j ($j = 1, \ldots, m$) are all the simple, connected, normal Lie subgroups of G, and K is a finite subgroup of the centre of the product. As a corollary, a similar structure theorem is given in which the G_j are also simply connected. This latter result is then generalised to arbitrary compact connected groups.

The decision on whether to include a particular result was based almost entirely on whether or not it was required for the proofs of the above structure theorems. This procedure accounted for the inclusion of most of the fundamental results and concepts in the theory of Lie groups; to round off the notes it only remained to add a few divertimenti such as the contents of Chapter 4 on the geometry of Lie groups or the list in Chapter 6 of necessary and sufficient conditions for a compact group to be Lie.

Chapter 1 contains results in the theory of analytic manifolds which are basic to the study of Lie groups. Chapter 2 begins the study of Lie groups and it is here that most of the fundamental concepts such as Lie algebras, left invariant vector fields, 1-parameter subgroups and the exponential map are introduced. In Chapter 3 the first deep result is presented; this is the Campbell-Baker-Hausdorff formula and it describes a relationship between the group structure of a Lie group and the algebraic structure of its Lie algebra. Chapter 4 introduces the notion of a geodesic on a Lie group and uses the resulting ideas to show that the exponential map is surjective whenever the Lie group is compact and connected. The correspondence between Lie subgroups of a Lie group and subalgebras of its Lie algebra is treated in Chapter 5. Chapter 6 begins with a list of conditions which are necessary and sufficient for a compact group to be Lie and ends with the structure results mentioned above. An appendix contains all the results on locally compact topological groups and their representations used in the body of the notes.

Further remarks, historical and motivational, on the contents of a chapter are given at the end of that chapter, along with related exercises. That a piece of theory is essential to a particular proof is no bar to it being included as an exercise if it is fairly straightforward or if it is fully treated in the literature.

I gave a course on some of the topics treated in these notes during 1973 at the Australian National University to an audience consisting mainly of postgraduate students, and then in 1974 at the University of New South Wales. These notes derive from these courses and in particular from duplicated notes of the earlier chapters. I am grateful to the people attending these courses for improvements of a number of arguments and in particular to Dr. Graham Wood for his reading of Chapter 1 and subsequent discussions. It was he who developed the local coordinate-free formula given in 1.3.2 and 1.3.3 for the Lie product of two analytic vector fields.

Finally, I feel that this preface would not be complete without some mention of the role of diagrams. Even though a large number of the concepts and results of manifolds and Lie groups have a strong pictorial or diagrammatic aspect, my experience is that diagrams in mathematics books are often of little value without a personal explanation. For this reason and because of widely varying preferences as to style, apart from several 'commutative arrow diagrams', none have been included here. However, without doubt they are valuable in developing an intuition in this area and the reader is strongly encouraged to experiment with them. Also some have found benefit in reformulating key results in terms of coordinates.

Kensington, 1976 J. F. P.

1·Analytic manifolds

This chapter contains the basic theory of analytic manifolds modelled on finite-dimensional real vector spaces. As promised, a coordinate-free approach will be used with emphasis on global definitions and properties. One of the reasons for including this chapter, instead of referring the reader to one or other of the numerous texts on manifolds, is to allow the reader to gain familiarity with this approach since it will permeate our whole treatment of Lie groups. Once one does away with coordinates it becomes obvious that large chunks of the theory of manifolds can be effortlessly generalised to manifolds modelled on infinite-dimensional spaces. We will have no need here of this degree of generality for reasons explained in the Notes at the end of the chapter, but the interested reader should consult the works of Lang, [1] and [2]. Since the theory of manifolds is one of the three legs on which the study of Lie groups stands, the other two being the theory of locally compact groups and the theory of Lie algebras, it is important that the ideas in this chapter, few though they may be, are well understood.

1.1 Manifolds and differentiability

1.1.1 **Manifolds.** Let M be a nonvoid Hausdorff topological space and E a real finite-dimensional vector space. If $\phi : U \to V$ is a homeomorphism between open subsets U and V of E and M respectively, then we say that ϕ is a <u>chart on</u> M. Also, if $p \in V$, then we say that ϕ is a <u>chart about</u> p. (Thanks to the infiltration of notions from category theory, it is now respectable to suppose that whenever a function is specified, its domain and codomain are automatically specified along with it. Hence there is no need to always explicitly write each function as a triple. We will adopt this convention here and immediately make use of it by supposing that whenever ϕ, ϕ_α and ϕ_β are charts, then their domains are U, U_α and U_β respectively, and their codomains (which in this case are also their ranges) are V, V_α and V_β respectively, unless otherwise specified.)

Suppose that ϕ_α is a chart on M for each α in some index set A. Then this collection, denoted $(\phi_\alpha : \alpha \in A)$, is called an atlas on M provided:

(i) each U_α is contained in the same finite-dimensional space, E say; and

(ii) the union of the V_αs is equal to M.

In this case we say that M is a manifold or M is a manifold modelled on E. (When we wish to be completely explicit we will say that the pair $(M, (\phi_\alpha : \alpha \in A))$ is a manifold. However, when no confusion seems possible, we will write only 'M is a manifold'.) The dimension of M as a manifold is defined as the dimension of E. Regarding the invariance of dimension, see Exercise 1.C(i).

It is obvious that every open subset U of a real finite-dimensional space E is a manifold when equipped with its identity map i. Henceforth, whenever we refer to such a set U as a manifold, its atlas will always be assumed to be $i : U \to U$. Less trivial examples of a manifold will be given in Chapter 2 after the definition of a Lie group.

Convention. It is easily seen that a Hausdorff topological space M can be equipped with a 0-dimensional atlas if and only if the topology of M is discrete. Thus, even though all the ensuing results on manifolds and Lie groups are valid for the 0-dimensional case, they are banal. Hence we will make the convention that the dimensions of all linear spaces, manifolds, and Lie groups are at least 1. In those cases when we want to emphasise that the dimension of a linear real space is n, we will often write it as $\mathbf{R}^n$, where $\mathbf{R}$ denotes the real line. Generally, however, such finite-dimensional real spaces will be denoted by E or F.

1.1.2 **Differentiable maps.** The abstract definition of the derivative of a map between finite-dimensional vector spaces is the main point of departure from the classical approach to differentiable manifolds to one involving no explicit mention of coordinates. Given a function f from an open subset U of a finite-dimensional real space E into another such space F, then we say that f is differentiable at x in U if there exists a linear map $f'(x) : E \to F$ such that

$$(1.1.1) \qquad \lim_{\varepsilon \to 0} \varepsilon^{-1}(f(x + \varepsilon h) - f(x)) = f'(x)h$$

uniformly for h in any bounded subset of U (provided, of course, that $x + \varepsilon h \in U$). This is readily seen to be equivalent to the existence of a linear map $f'(x) : E \to F$ such that

$$(1.1.2) \qquad \lim_{h \to 0, h \neq 0} \frac{\|f(x+h) - f(x) - f'(x)h\|}{\|h\|} = 0 .$$

Here the norm is taken as any one of the equivalent norms which make the finite-dimensional space E into a Banach space. (See Edwards [1, Proposition 1.9.6].) Throughout the sequel, whenever the need for a topology on a finite-dimensional vector space arises, then it will always be taken to be the topology induced from such a norm.

Exercise 1.A collects together some of the elementary properties of this derivative, for example, the uniqueness of the linear map $f'(x)$.

If f is differentiable at each point of U, we say that f is differentiable on U. In this case we have the function

$$f' : U \to \hom(E, F),$$

where $f' : x \mapsto f'(x)$ and $\hom(E, F)$ is the linear space of linear maps from E into F. Continuing in this way it is clear that we may have higher order derivatives $f'' = (f')'$, $f''' = (f'')'$, and so on. In this way we arrive at the notion of a smooth function (at a point or on an open set) being a function which possesses derivatives of all orders (in a neighbourhood of the point or in the open set). Suppose that f is as above and that f'' exists on U, then f'' is a function from U into $\hom(E, \hom(E, F))$. As is customary, we identify this latter space in the canonical manner with $\hom^2(E \times E, F)$, the bilinear maps from $E \times E$ into F. In fact, throughout we adopt the convention that if $f^{(p)}$ exists on U, then its image space is $\hom^p(E \times \ldots \times E$ (p times), $F)$. This simplifies a number of expressions, including Taylor's expansion in 1.1.5 below.

If E, F and G are real finite-dimensional spaces and $f : E \to F$ and $g : F \to G$ are differentiable at x and $f(x)$ respectively, a classical result (included below in Exercise 1. A) states that $g \circ f$ is also differentiable at x and moreover:

$$(1.1.3) \qquad (g \circ f)'(x) = g'(f(x)) \circ f'(x).$$

1.1.3 **Remarks.** The notion of the derivative given above is often called the Fréchet derivative. For Banach spaces the study of this derivative forms Chapter VIII of Dieudonné [1], while Averbukh and Smolyanov [1] study this and related derivatives on topological vector spaces in general. For example, these latter authors show that in a certain sense the Fréchet derivative is the weakest type of differentiation for which the first order chain rule, formula (1.1.3) above, is valid for finite-dimensional spaces [1, p. 74].

1.1.4 **Maps from R.** When considering a differentiable map $f : \mathbf{R} \to E$, then $f'(x)$ satisfies

$$f'(x)(t) = f'(x)(1).t \quad \text{for each } t \text{ in } \mathbf{R}.$$

Thus $f'(x)$ is completely described by its value at 1 and we often write $f'(x)$ in place of $f'(x)(1)$. (This is precisely what happens in the classical case of functions from $\mathbf{R}$ into $\mathbf{R}$ where the derivative $f'(x)$ is taken to be a number as opposed to an operator.)

1.1.5 **Analytic functions.** Suppose that f is a smooth function from an open subset U of a real finite-dimensional space E into another such space F. Let x in U and y in E be such that $x + ty \in U$ for all $t \in [0, 1]$. If $y^{(m)}$ denotes the m-tuple $(y, \ldots, y)$, then

$$(1.1.4) \qquad f(x+y) = f(x) + \frac{1}{1!} f'(x)y + \ldots + \frac{1}{m!} f^{(m)}(x)y^{(m)} + R_{m+1}(y)$$

for each $m \in Z^+ = \{0, 1, 2, \ldots\}$, where the error term R_{m+1} satisfies $\lim_{y \to 0} R_{m+1}(y).\|y\|^{-m} = 0$. (See Exercise 1.D, where one particular version of the error term is described.) The sum (1.1.4) is called Taylor's formula of degree m.

Just as in the 1-dimensional case, we say that a smooth function $f : U \to F$ is (real) analytic on U if for each x in U there exists an open ball $B \subseteq U$ with centre x such that for all $z = x + y$ in B, the series

$$(1.1.5) \qquad \sum_{m=0}^{\infty} \frac{1}{m!} f^{(m)}(x)y^{(m)}$$

is absolutely convergent (that is, $\sum_m \frac{1}{m!} \| f^{(m)}(x)y^{(m)} \|$ is convergent, where the norm is that of F) and converges to $f(z)$. The function $f : U \to F$ is said to be analytic at x if it is analytic in some neighbourhood of x.

Examples. (i) If $f : E \to F$ and $g : F \to G$ are analytic at x and $f(x)$ respectively, then $g \circ f$ is analytic at x.

(ii) If $f : U \to F$ is analytic on U, an open subset of E, then $f^{(k)}$ is also analytic on U for each $k \in Z^+$ and its expansion at $x + y \in U$, where $x \in U$, is $\sum_m \frac{1}{m!} f^{(k+m)}(x)y^{(m)}$; in other words,

$$f^{(k)}(x+y)(u_1, \ldots, u_k) = \sum_{m=0}^{\infty} \frac{1}{m!} f^{(k+m)}(x)(\overbrace{y, \ldots, y}^{m \text{ terms}}, u_1, \ldots, u_k).$$

The validation of these two examples is left as an exercise for the interested reader.

(iii) Examples of smooth functions which are not analytic are well known. Even the absolute convergence of (1.1.5) in a ball is not sufficient to ensure the analyticity of the function at the centre of the ball concerned. For example, consider the smooth function $g : \mathbf{R} \to \mathbf{R}$ defined by

$$g(x) = \begin{cases} e^{-1/x^2} & \text{for } x \neq 0 \\ 0 & \text{for } x = 0 \end{cases} .$$

It satisfies $g^{(m)}(0) = 0$ for all $m \in Z^+$ so that the series (1.1.5) is absolutely convergent for all $y \in \mathbf{R}$, but it only converges to $g(x)$ when $x = 0$.

1.1.6 **Smooth atlases and manifolds.** An atlas $(\phi_\alpha : \alpha \in A)$ on the Hausdorff topological space M is said to be smooth if each of the functions $\phi_\beta^{-1} \circ \phi_\alpha$ is smooth on $\phi_\alpha^{-1}(V_\alpha \cap V_\beta)$. Such an atlas is said to be maximal if whenever U and V are open subsets of E and M respectively and $\phi : U \to V$ is a homeomorphism with the property that the functions

$$(1.1.6) \quad \phi_\alpha^{-1} \circ \phi : \phi^{-1}(V \cap V_\alpha) \to \phi_\alpha^{-1}(V \cap V_\alpha)$$

$$(1.1.7) \quad \phi^{-1} \circ \phi_\alpha : \phi_\alpha^{-1}(V \cap V_\alpha) \to \phi^{-1}(V \cap V_\alpha)$$

are smooth for each $\alpha \in A$, then $\phi \in (\phi_\alpha : \alpha \in A)$.

Lemma. Every smooth atlas on M is contained in a unique maximal smooth atlas.

Proof. If $(\phi_\alpha : \alpha \in A)$ is smooth on M, let $(\phi_{\alpha'} : \alpha' \in A')$ denote the collection of all maps ψ which are homeomorphisms between open subsets of E and M and which satisfy (1.1.6) and (1.1.7). This collection is an atlas with the desired properties. //

As a matter of terminology, a smooth atlas is said to generate the unique maximal smooth atlas which contains it.

Definition. A manifold $(M, (\phi_\alpha : \alpha \in A))$ is said to be smooth if the atlas $(\phi_\alpha : \alpha \in A)$ is both smooth and maximal.

In practice it is more usual to work with generating atlases rather than the corresponding maximal atlases since, as for the case of sub-bases in topology, most of the properties with which we are concerned are valid on a maximal atlas if valid on any of its generating atlases. Thus if we specify a smooth atlas $(\phi_\alpha : \alpha \in A)$ on M and then refer to $(M, (\phi_\alpha : \alpha \in A))$ as a smooth manifold, the precise meaning is that we are to take M equipped with the maximal smooth atlas generated by $(\phi_\alpha : \alpha \in A)$.

For example, if M is a real finite dimensional space equipped with its usual topology and $i : M \to M$ is the identity map, then $(M, \{i\})$ is a smooth manifold. This is simple enough but even here the maximal smooth atlas generated by i contains a superabundance of members. As an exercise describe them.

1.1.7 **Smooth maps.** If M and N are smooth manifolds with smooth atlases $(\phi_\alpha : \alpha \in A)$ and $(\psi_\beta : \beta \in B)$ respectively, we say that a map f from U, an open subset of M, into N is smooth if each of the maps $\psi_\beta^{-1} f \phi_\alpha$, defined on $\phi_\alpha^{-1}(V_\alpha \cap f^{-1}(V_\beta))$, is smooth.

In particular, f is smooth at the point x if and only if

(i) $\psi_\beta^{-1} f \phi_\alpha$ is smooth at $\phi_\alpha^{-1}(x)$ for every pair $(\phi_\alpha, \psi_\beta)$ satisfying

(ii) $x \in \operatorname{codom} \phi_\alpha$, $f(x) \in \operatorname{codom} \psi_\beta$.

However, because of the smoothness of the atlases involved, we need only consider the smoothness of (i) for any particular pair satisfying (ii). For example, suppose that (i) is satisfied by the pair $(\phi_\alpha, \psi_\beta)$ in (ii), and further suppose that $(\phi_{\alpha'}, \psi_{\beta'})$ is another pair satisfying (ii). Then certainly $\phi_\alpha^{-1}\phi_{\alpha'}$ and $\psi_{\beta'}^{-1}\psi_\beta$ are smooth at $\phi_{\alpha'}^{-1}(x)$ and $\psi_\beta^{-1}(f(x))$ respectively. Thus

$$\psi_{\beta'}^{-1} f \phi_{\alpha'} = (\psi_{\beta'}^{-1}\psi_\beta)(\psi_\beta^{-1} f \phi_\alpha)(\phi_\alpha^{-1}\phi_{\alpha'})$$

is smooth at $\phi_{\alpha'}^{-1}(x)$ showing that (i) is also satisfied by the pair $(\phi_{\alpha'}, \psi_{\beta'})$.

If M and N are smooth manifolds and if $f : M \to N$ is a homeomorphism such that f and f^{-1} are smooth, then f is said to be a diffeomorphism.

1.1.8 **Analytic manifolds.** If in the above definitions of smooth atlases, manifolds and maps between manifolds, we replace 'smooth' by 'analytic', then we arrive at the definitions of analytic atlases, manifolds and maps between manifolds. If a homeomorphism and its inverse are analytic, then the homeomorphism is said to be an analytic homeomorphism or even an analytic diffeomorphism.

1.1.9 **A condition for analyticity.** Let M and N be analytic manifolds with analytic atlases $(\phi_\alpha : \alpha \in A)$ and $(\psi_\beta : \beta \in B)$ respectively. Analogously to 1.1.7, a map $f : M \to N$ is analytic at x in M if and only if there exist charts ϕ_α about x and ψ_β about f(x) such that $\psi_\beta^{-1} f \phi_\alpha$ is analytic at x.

1.2 The tangent bundle

1.2.1 The basic idea. Let U be an open subset of a finite-dimensional real space E. If $\xi : (-\varepsilon, \varepsilon) \to U$, $\varepsilon > 0$, is an analytic curve satisfying $\xi(0) = p$, then, either by calculus or imagination, ξ has a tangent at p. Moreover, two analytic curves passing through p have the same tangent provided they have the same 'direction' and the same 'speed' at p. In mathematical terms, the tangent to ξ at p is defined as $\xi'(0)$ (or, $\xi'(0)(1)$), a vector in E. Thus the tangent space of U at p may be thought of as E and can be given a concrete realisation as the family of analytic curves, equivalent modulo their derivatives at p, passing through p.

Now suppose that M is an analytic manifold modelled on E. Let $\mathcal{C}_p(M)$ denote the set of analytic maps ξ from open neighbourhoods of 0 in $\mathbf{R}$ into M which satisfy $\xi(0) = p$. The derivative of ξ is not defined, but we circumvent this difficulty by considering the derivative of $\phi_\alpha^{-1}\xi$ at 0, where ϕ_α is a chart about p. Define an equivalence relation on $\mathcal{C}_p(M)$ by

$$(1.2.1) \quad \xi \sim \eta \text{ if } (\phi_\alpha^{-1}\xi)'(0) = (\phi_\alpha^{-1}\eta)'(0).$$

Let $[[\xi]]_p$ denote the class of curves equivalent to ξ (and note that if two curves are equivalent as in (1.2.1), then they are equivalent for each chart about p); the <u>tangent space</u> to M at p is defined as the set of all such equivalence classes.

Proceeding in the opposite direction, given any v in E and any analytic chart ϕ_α about p, can we always find a curve ξ in $\mathcal{C}_p(M)$ such that $(\phi_\alpha^{-1}\xi)'(0) = v$? The answer is 'yes' and forms part of Exercise 1.B.

Thus this fairly intuitive approach to the idea of a tangent space at a point of a manifold modelled on E shows that it is always isomorphic to E. Once this point is seen, it becomes notationally easier to proceed straight to E without any consideration of analytic curves. This will be done in the next subsection. A further approach to tangent spaces via 'derivation mappings' is contained in Exercise 1. F.

1. 2. 2 Tangent spaces. Suppose that $(M, (\phi_\alpha : \alpha \in A))$ is an analytic manifold and that p is a point on M. We define an equivalence relation on the pairs (ϕ_α, v), where ϕ_α is a chart about p and $v \in E$ by

$$(1.2.2) \quad (\phi_\alpha, v) \sim_p (\phi_\beta, w) \text{ if } (\phi_\beta^{-1}\phi_\alpha)'(\phi_\alpha^{-1}p)(v) = w.$$

This process may be thought of in the following way: corresponding to each chart about p we attach the space E to the point p, each pair of attachments then being identified under the equivalence relation (1. 2. 2). This, in effect, leaves us an object at p independent of all charts about p. We will see shortly that this object has a natural linear structure which makes it isomorphic with E. With this structure it is called the tangent space of M at p. Motivation for (1. 2. 2) comes by pursuing further the preceding discussion in 1. 2. 1 on analytic curves ξ on M with $\xi(0) = p$. It is reasonable to ask that the tangent at p corresponding to ξ depends on ξ only and not on different charts ϕ_α and ϕ_β, say, about p, so in some sense we want $(\phi_\alpha^{-1}\xi)'(0)$ and $(\phi_\beta^{-1}\xi)'(0)$ to be equivalent. They are, under (1. 2. 2)!

Before going further, we should verify that (1. 2. 2) is indeed an equivalence relation. To do this we must ascertain that it is (i) reflective, (ii) symmetric and (iii) transitive.

(i) It is immediate that $(\phi_\alpha, v) \sim_p (\phi_\alpha, v)$ and hence that the relation is reflexive.

(ii) Suppose that $(\phi_\alpha, v) \sim_p (\phi_\beta, w)$ and hence $(\phi_\beta^{-1}\phi_\alpha)'(\phi_\alpha^{-1}p)v = w$; the relation is symmetric provided $(\phi_\alpha^{-1}\phi_\beta)'(\phi_\beta^{-1}p)w = v$. Now

$$\begin{aligned}(\phi_\alpha^{-1}\phi_\beta)'(\phi_\beta^{-1}p)w &= (\phi_\alpha^{-1}\phi_\beta)'(\phi_\beta^{-1}p)(\phi_\beta^{-1}\phi_\alpha)'(\phi_\alpha^{-1}p)v \\ &= (\phi_\alpha^{-1}\phi_\beta\phi_\beta^{-1}\phi_\alpha)'(\phi_\alpha^{-1}p)v\end{aligned}$$

(using the chain rule (1. 1. 3))

$$= (\mathrm{Id})'(\phi_\alpha^{-1}p)v$$

$$= v,$$

as required.

(iii) If $(\phi_\alpha, v) \sim_p (\phi_\beta, w)$ and $(\phi_\beta, w) \sim_p (\phi_\gamma, u)$, then the transitivity of the relation is established provided $(\phi_\alpha, v) \sim_p (\phi_\gamma, u)$, that is, provided $(\phi_\gamma^{-1}\phi_\alpha)'(\phi_\alpha^{-1}p)v = u$. Making use of the chain rule yields

$$\begin{aligned}(\phi_\gamma^{-1}\phi_\alpha)'(\phi_\alpha^{-1}p)v &= (\phi_\gamma^{-1}\phi_\beta\phi_\beta^{-1}\phi_\alpha)'(\phi_\alpha^{-1}p)v \\ &= (\phi_\gamma^{-1}\phi_\beta)'(\phi_\beta^{-1}p)(\phi_\beta^{-1}\phi_\alpha)'(\phi_\alpha^{-1}p)v \\ &= (\phi_\gamma^{-1}\phi_\beta)'(\phi_\beta^{-1}p)w \\ &= u,\end{aligned}$$

the sought-after relation.

Denote the class equivalent to (ϕ, v) at p by $[\phi, v]_p$. The set of these equivalence classes is given a linear structure by defining

$$\lambda[\phi_\alpha, v]_p = [\phi_\alpha, \lambda v]_p \quad \text{for } \lambda \text{ in } \mathbf{R}, \text{ and}$$
$$[\phi_\alpha, v]_p + [\phi_\beta, w]_p = [\phi_\beta, (\phi_\beta^{-1}\phi_\alpha)'(\phi_\alpha^{-1}p)v + w]_p.$$

(For complete rigour, it should be checked that these linear operations are well-defined but we will take the easy way and refer the reader to Exercise 1. B.)

The set of all equivalence classes under (1. 2. 2) with the linear structure defined above will be denoted by $T_p(M)$ and is called the tangent space of M at p. Clearly it is linearly isomorphic with E. Since, in effect, we are adjoining a copy of E to each point of M, if p and q are distinct points in M, the question of whether $T_p(M)$ and $T_q(M)$ overlap does not arise.

1. 2. 3 **The tangent bundle.** Suppose that $(M, (\phi_\alpha : \alpha \in A))$ is an analytic manifold modelled on E; the tangent bundle of M is defined as the (disjoint) union of the tangent spaces $T_p(M)$, p running over M, and will be denoted by $T(M)$.

The next proposition shows that T(M) can always be given, in a natural manner, a Hausdorff topology and an analytic atlas under which it becomes an analytic manifold modelled on $E \times E$. In the sequel T(M) will always be assumed to have this extra structure. Define a map $\pi : T(M) \to M$, called the natural projection, by $\pi([\phi_\alpha, v]_p) = p$. Corresponding to each α in A define

$$\tau_\alpha : U_\alpha \times E \to \pi^{-1}(V_\alpha) \text{ by } \tau_\alpha(u, v) = [\phi_\alpha, v]_{\phi_\alpha u}.$$

Notice that τ_α is a bijection and that the union of the codomains of the τ_α is equal to T(M). Suppose that for α, β in A the codomains of τ_α and τ_β overlap, that is, that

$$\pi^{-1}(V_\alpha) \cap \pi^{-1}(V_\beta) \neq \emptyset.$$

Since $\pi^{-1}(V_\alpha) \cap \pi^{-1}(V_\beta) = \pi^{-1}(V_\alpha \cap V_\beta)$, we have $V_\alpha \cap V_\beta \neq \emptyset$. Let (u, v) belong to $\tau_\alpha^{-1}(\pi^{-1}V_\alpha \cap \pi^{-1}V_\beta)$; then

$$\begin{aligned}\tau_\beta^{-1}\tau_\alpha(u, v) &= \tau_\beta^{-1}([\phi_\alpha, v]_{\phi_\alpha u})\\ &= \tau_\beta^{-1}([\phi_\beta, (\phi_\beta^{-1}\phi_\alpha)'(u)v]_{\phi_\beta(\phi_\beta^{-1}\phi_\alpha u)})\\ &= (\phi_\beta^{-1}\phi_\alpha u, (\phi_\beta^{-1}\phi_\alpha)'(u)v).\end{aligned}$$

Now

$$\begin{aligned}\tau_\alpha^{-1}(\pi^{-1}V_\alpha \cap \pi^{-1}V_\beta) &= \tau_\alpha^{-1}\pi^{-1}(V_\alpha \cap V_\beta)\\ &= \phi_\alpha^{-1}(V_\alpha \cap V_\beta) \times E\end{aligned}$$

and so the analyticity of $\tau_\beta^{-1}\tau_\alpha$ on $\tau_\alpha^{-1}(\pi^{-1}V_\alpha \cap \pi^{-1}V_\beta)$ follows from the analyticity of $\phi_\beta^{-1}\phi_\alpha$ on $\phi_\alpha^{-1}(V_\alpha \cap V_\beta)$.

The collection of open sets $\tau_\alpha(U \times V)$, α ranging over A, U ranging over open subsets of U_α, and V ranging over open subsets of E, may easily be seen to form the basis of a Hausdorff topology on T(M) ... see Exercise 1.E ... and under this topology we have just shown that:

Proposition. $(T(M), (\tau_\alpha : \alpha \in A))$ is an analytic manifold modelled on $E \times E$.

Each chart τ_α is often referred to as a trivializing chart of $T(M)$ and it shows that the tangent bundle over V_α can be identified with $U_\alpha \times E$.

1.2.4 **Manifold derivatives.** If $f : M \to N$ is an analytic map between analytic manifolds $(M, (\phi_\alpha))$ and $(N, (\psi_\beta))$ and $p \in M$, then we define a linear map $f_{*,p} : T_p(M) \to T_{f(p)}(N)$, called the manifold derivative of f at p, by

$$f_{*,p} : [\phi_\alpha, v]_p \mapsto [\psi_\beta, (\psi_\beta^{-1} f \phi_\alpha)'(\phi_\alpha^{-1} p)v]_{f(p)} .$$

To see that this function is well-defined it must be shown that $(\psi_\beta, (\psi_\beta^{-1} f \phi_\alpha)'(\phi_\alpha^{-1} p)v) \sim_{f(p)} (\psi_\delta, (\psi_\delta^{-1} f \phi_\gamma)'(\phi_\gamma^{-1} p)w)$ whenever $(\phi_\alpha, v) \sim_p (\phi_\gamma, w)$. This is easily accomplished by two applications of the chain rule for first derivatives to definition (1.2.2) of the equivalence relation.

By letting p range over M we define a tangent bundle map $f_* : T(M) \to T(N)$ by $f_* = f_{*,p}$ on $T_p(M)$.

Suppose that both $f : M \to N$ and $g : N \to P$ are analytic maps, where M, N and P are analytic manifolds. Then the first order chain rule (1.1.3) allows us to write

$$(g \circ f)_{*,p} = g_{*,f(p)} \circ f_{*,p}$$

and hence

$$(1.2.3) \quad (g \circ f)_* = g_* \circ f_*.$$

It will be of benefit later to introduce now the notions of rank and immersion.

1.2.5 **Rank.** If $f : M \to N$ is an analytic map, where M and N are analytic manifolds, then the rank of f at p is defined as the dimension of the image of $f_{*,p}$ in $T_{f(p)}(N)$.

Lemma. Denote the kernel and the image of $f_{*,p}$ by $\mathrm{Ker}(f_{*,p})$ and $\mathrm{Im}(f_{*,p})$ respectively. Then

$$\dim(M) = \dim(\mathrm{Ker}\, f_{*,p}) + \dim(\mathrm{Im}\, f_{*,p}).$$

Proof. Elementary. From a more advanced standpoint this lemma amounts to saying that any short exact sequence of finite-dimensional vector spaces splits. //

Corollary. The rank of $f : M \to N$ at p is at most the smaller of $\dim(M)$ and $\dim(N)$.

1.2.6 **Immersion.** An analytic map $f : M \to N$ is called an immersion if its rank is equal to the dimension of M at all points of M. In other words, an immersion is an analytic map f such that $f_{*,p}$ is one-to-one for each p in M.

1.3 Vector fields

Throughout this subsection we suppose that $(M, (\Phi_\alpha : \alpha \in A))$ is an analytic manifold modelled on the finite-dimensional real space E. A map X from M into $T(M)$ is called a vector field provided $\pi X(p) = p$ for all p in M. Since M and $T(M)$ are both analytic manifolds we may also talk about analytic vector fields.

A useful way of visualising a vector field is to first think of the tangent space $T_p(M)$ of M at p as a 'fibre' growing out of the point p, $T(M)$ thus becoming the 'bundle' of all such fibres as p ranges over M. In this picture a vector field is a process of associating with each point of M a point in its fibre. In other words, a vector field is a cross-section of the tangent bundle. (We have used the suggestive terms 'fibre' and 'bundle' since tangent bundles are a special case of the general notion of a 'fibre bundle'. For details of the general case see Steenrod [1], and for applications to infinite-dimensional manifolds see Eells [1] and Lang [2].)

We now show that the set of analytic vector fields on M can be given both a linear and an algebraic structure.

1.3.1 **Linear structure.** If X and Y are analytic vector fields on M and $\lambda \in R$, then we define λX and $X + Y$ as the operators

$$\lambda X : p \mapsto \lambda X(p) \quad \text{and} \quad X + Y : p \mapsto X(p) + Y(p)$$

respectively, where the images are calculated by using the linear-space operations on $T_p(M)$ defined in 1.2.2. Then both λX and $X + Y$ are also analytic vector fields.

1.3.2 **Algebraic structure.** There are a number of equivalent ways of defining the 'Lie' product of two analytic vector fields. We will give a fairly cumbersome, but direct, definition of the product followed by an equivalent formulation which will motivate the introduction of such a product. For each α in A, define the projection $\rho_\alpha : T_p(M) \to E$ by $\rho_\alpha([\phi_\alpha, w]_p) = w$.

Definition. Given two analytic vector fields X, Y on M, we define their product $[X, Y] : M \to T(M)$ by;

$$[X, Y](p) = [\phi_\alpha, (\rho_\alpha Y\phi_\alpha)'(v)(\rho_\alpha X\phi_\alpha(v)) - (\rho_\alpha X\phi_\alpha)'(v)(\rho_\alpha Y\phi_\alpha(v))]_p,$$

where $v = \phi_\alpha^{-1}(p)$.

We must check (i) that the map is independent of the particular chart ϕ_α, and (ii) that it is analytic.

(i) Suppose that $\phi_\alpha(v) = \phi_\beta(w) = p$; we require

$$[\phi_\alpha, (\rho_\alpha Y\phi_\alpha)'(v)(\rho_\alpha X\phi_\alpha(v)) - (\rho_\alpha X\phi_\alpha)'(v)(\rho_\alpha Y\phi_\alpha(v))]_p$$

to be equivalent to

$$[\phi_\beta, (\rho_\beta Y\phi_\beta)'(w)(\rho_\beta X\phi_\beta(w)) - (\rho_\beta X\phi_\beta)'(w)(\rho_\beta Y\phi_\beta(w))]_p.$$

That is

$$(\phi_\beta^{-1}\phi_\alpha)'(\phi_\alpha^{-1}p)[(\rho_\alpha Y\phi_\alpha)'(v)(\rho_\alpha X\phi_\alpha(v)) - \ldots]$$
$$= (\rho_\beta Y\phi_\beta)'(w)(\rho_\beta X\phi_\beta(w)) - \ldots\ .$$

Now $v = \phi_\alpha^{-1}\phi_\beta(w)$ and it is easily seen that $\rho_\alpha = (\phi_\alpha^{-1}\phi_\beta)'(w)\rho_\beta$, so that the left hand side of the above becomes

$$(\phi_\beta^{-1}\phi_\alpha)'(\phi_\alpha^{-1}\phi_\beta w) \circ [((\phi_\alpha^{-1}\phi_\beta)'(w)\rho_\beta Y\phi_\beta\phi_\beta^{-1}\phi_\alpha)'(\phi_\alpha^{-1}\phi_\beta w)$$

$$((\phi_\alpha^{-1}\phi_\beta)'(w)\rho_\beta X\phi_\beta(w)) - \dots]$$

$$= (\phi_\beta^{-1}\phi_\alpha)'(\phi_\alpha^{-1}\phi_\beta w) \circ [(\phi_\alpha^{-1}\phi_\beta)'(w)(\rho_\beta Y\phi_\beta)'(w)(\phi_\beta^{-1}\phi_\alpha)'(\phi_\alpha^{-1}\phi_\beta w) \circ$$

$$(\phi_\alpha^{-1}\phi_\beta)'(w)\rho_\beta X\phi_\beta(w) - \dots]$$

$$= (\rho_\beta Y\phi_\beta)'(w)(\rho_\beta X\phi_\beta(w)) - \dots,$$

as required, since $(\phi_\beta^{-1}\phi_\alpha)'(\phi_\alpha^{-1}\phi_\beta w) \circ (\phi_\alpha^{-1}\phi_\beta)'(w) = I$, the identity operator.

(ii) To check the analyticity of $[X, Y]$ at the point p, from 1.1.9 we know that only the analyticity of

$$v \mapsto \tau_\alpha^{-1}[X, Y]\phi_\alpha(v)$$

need be determined at $v = \phi_\alpha^{-1}(p)$. The assumption that X is analytic at p means precisely that

$$v \mapsto \tau_\alpha^{-1}X\phi_\alpha(v) = (v, \rho_\alpha X\phi_\alpha(v))$$

is analytic at $\phi_\alpha^{-1}(p)$, and similarly for Y. Now

$$\tau_\alpha^{-1}[X, Y]\phi_\alpha(v) = (v, (\rho_\alpha Y\phi_\alpha)'(v)(\rho_\alpha X\phi_\alpha)(v) - \dots)$$

by definition, and so the analyticity of $\tau_\alpha^{-1}[X, Y]\phi_\alpha$ (and hence of $[X, Y]$) follows from the analyticity of X and Y and the fact that $x \mapsto g'(x)f(x)$ is analytic at $a \in E$ whenever this property is shared by both $g, f : E \to E$.

Example. If M is an open subset of E, then we are always within the smooth (trivial) atlas i and so no confusion will arise if v and $\rho_i[i, v]_p$ are identified. If we also suppress all direct mention of i, then the bracket product of smooth vector fields X and Y is simply expressed as

$$(1.3.1)\quad [X, Y]_p = Y'(p)X(p) - X'(p)Y(p).$$

Returning to the general case, since we know that both M and T(M) are locally homeomorphic to open subsets of real finite-dimensional spaces, then it is reasonable to say that (1.3.1) is the local representation of [X, Y].

1.3.3 **Derivations.** Let $\mathcal{F}_\alpha$ denote the set of analytic real-valued functions on some V_α, where $(M, (\phi_\alpha : \alpha \in A))$ is an analytic manifold. For each analytic vector field X, define a linear operator $\tilde{X} : \mathcal{F}_\alpha \to \mathcal{F}_\alpha$ by

$$(1.3.2)\quad \tilde{X}f : p \mapsto \rho f_{*,p} X(p) \quad \text{for } p \in V_\alpha,$$

where $\rho = \rho(f, p)$ is the natural projection from $T_{f(p)}(R)$ onto R given by $\rho([i, w]_{f(p)}) = w$. In practice $T_{f(p)}$ and R will be identified so that

$$(1.3.2)'\quad (\tilde{X}f)(p) = f_{*,p} X(p) \quad \text{for } p \in V_\alpha,\ f \in \mathcal{F}_\alpha.$$

(Strictly speaking the domain of $f_{*,p}$ is $T_p(V_\alpha)$, but this is canonically isomorphic with $T_p(M)$.)

The idea behind this definition is that it is an abstraction of the classical notion of a directional derivative; $\tilde{X}f(p)$ is the 'directional derivative' of f at p with respect to the tangent vector X(p).

The operator $\tilde{X}$ clearly satisfies

$$(1.3.3)\quad \tilde{X}(fg)(p) = f(p)(\tilde{X}g)(p) + g(p)(\tilde{X}f)(p)$$

on V_α for $f, g \in \mathcal{F}_\alpha$. Any linear operator on $\mathcal{F}_\alpha$ which satisfies (1.3.3) is called a derivation. (Local derivations are defined in Exercise 1.F.)

Theorem. *If X and Y are analytic vector fields on an analytic manifold M, then* $[X, Y]^\sim = \tilde{X}\tilde{Y} - \tilde{Y}\tilde{X}$.

Proof. With ϕ_α, ρ_α and v as in the definition of the Lie product, $\tilde{X}f(p)$ can be written as

$$(f\phi_\alpha)'(v)(\rho_\alpha X\phi_\alpha(v)) = e \circ ((f\phi_\alpha)',\ \rho_\alpha X\phi_\alpha)(v),$$

where e is the canonical evaluation map from $\hom(E, R) \times E$ into R. Hence we may write $\tilde{X}(\tilde{Y}f)p$ locally as

$$(1.3.4) \quad \tilde{X}(\tilde{Y}f)p = [e \circ ((f\phi_\alpha)', \rho_\alpha Y\phi_\alpha)]'(v)(\rho_\alpha X\phi_\alpha(v)).$$

Now if E_1, E_2, E and F are real finite-dimensional spaces, $\theta : E_1 \times E_2 \to F$ is bilinear and $g : U \to E_1$, $h : U \to E_2$ are differentiable at $x \in U$, where U is open in E, then the function $P = \theta(g, h) : U \to F$ is differentiable at x with

$$(1.3.5) \quad P'(x)(u) = P(g'(x)u, h(x)) + P(g(x), h'(x)u)$$

for $u \in E$. Applying (1.3.5) to (1.3.4) results in

$$\tilde{X}(\tilde{Y}f)(p) = (f\phi_\alpha)''(v)(\rho_\alpha X\phi_\alpha(v))(\rho_\alpha Y\phi_\alpha(v)) + (f\phi_\alpha)'(v)(\rho_\alpha Y\phi_\alpha)'(v)(\rho_\alpha X\phi_\alpha)(v).$$

Now use the symmetry of the second derivative (Ex. 1.A) to show that the local expression for $\tilde{X}(\tilde{Y}f)p - \tilde{Y}(\tilde{X}f)p$ is precisely that of $\widetilde{[X, Y]}f(p)$. //

1.3.4 Corollary. The Lie product on the set of analytic vector fields is bilinear and satisfies

(i) $[X, X] = 0$, and

(ii) $[X, [Y, Z]] + [Y, [Z, X]] + [Z, [X, Y]] = 0$, for all analytic vector fields X, Y and Z.

Definition. A real linear space with a bilinear product satisfying (i) and (ii) above is said to be a Lie algebra. (A bilinear product satisfying (i) is said to be antisymmetric, while (ii) is usually referred to as Jacobi's identity. It follows from (i) and the bilinearity of the product that $[X, Y] = -[Y, X]$.)

Thus Corollary 1.3.4 states precisely that the set of analytic vector fields is a Lie algebra when it is equipped with its Lie product. In the next chapter we enter upon the study of Lie groups and it will quickly become apparent that a certain subalgebra of this Lie algebra contains a great deal of information about the underlying group. This subalgebra is called the Lie algebra of the underlying Lie group and the study of Lie groups essentially reduces to a study of the corresponding Lie algebras, each of which is finite-dimensional.

Let E denote a real finite-dimensional space. The exemplar of a Lie algebra is $\underline{\mathrm{gl}}(E)$, the space of all linear endomorphisms of E equipped with the product

$$[X, Y] = XY - YX.$$

Proof (of 1.3.4). The proof can be accomplished directly from the definition of the Lie product or via the theorem on derivations. In fact the result is an immediate consequence of the latter theorem once we show the following fact which is also of independent interest.

Fact. If $\tilde{X}f = 0$ for all f in $\mathcal{F}_\alpha$, then $X = 0$ on V_α.

Suppose that $X(p) = [\phi_\alpha, v]_p \neq 0$, in which case $v \neq 0$. Consequently there exists a linear functional $l : E \to \mathbf{R}$ such that $l(v) \neq 0$. Define f on V_α by $f = l\phi_\alpha^{-1}$; then f is analytic and hence is a member of $\mathcal{F}_\alpha$. Furthermore

$$(f\phi_\alpha)'(\phi_\alpha^{-1}p)(v) = l'(\phi_\alpha^{-1}p)(v) = l(v),$$

so that

$$\begin{aligned}\tilde{X}f(p) &= f_{*,p}[\phi_\alpha, v]_p \\ &= [i, (f\phi_\alpha)'(\phi_\alpha^{-1}p)v]_{f(p)} \\ &= [i, l(v)] \neq 0,\end{aligned}$$

and the proof is complete. //

Notes

There are two directions in which the concept of a smooth manifold can be naturally generalised.

(i) The classical direction is to vary the restrictions put on the maps $\phi_\beta^{-1}\phi_\alpha$, where ϕ_α and ϕ_β are charts. For example, the manifold $(M, (\phi_\alpha : \alpha \in A))$ is called a C^p-manifold if all the maps $\phi_\beta^{-1}\phi_\alpha$ possess p derivatives on $\phi_\alpha^{-1}(V_\alpha \cap V_\beta)$. However, when we come to consider compact Lie groups it will be seen that a manifold which is also a compact group (with the group and manifold topologies equivalent) is already an analytic manifold and that the group operations are analytic. (This is also known to be true for locally compact groups but it will not be proved here. For more details see the Notes after Chapter 6.) Thus any generalisation resulting from weaker differentiability conditions on the charts would be illusory once we enter the study of groups with a compatible manifold structure. All the results in the preceding chapter are valid mutatis mutandis for smooth manifolds in place of analytic manifolds. Also, with a little care, corresponding results can be formulated for C^p-manifolds.

(ii) The second direction in which to generalise the notion of a smooth manifold as described in this chapter is to model the manifold on an infinite-dimensional Banach space (even Fréchet space). (The definitions of a manifold and of smooth and analytic manifolds given in the beginning of this chapter remain meaningful when the associated real finite-dimensional space is replaced with a real infinite-dimensional Banach space.) Research in the field of infinite-dimensional manifolds has been very active over the last decade with the result that many diverse problems have been shown to have their bases in this area. See, for example, the survey article by Eells [1]. Also a book by Lang [2] shows clearly how numerous results, previously thought to belong to the realm of finite-dimensional manifolds, have natural analogues for manifolds modelled on Banach spaces and that happily the proofs are no more difficult.

Nevertheless, from our point of view (which is to present some of the basic results in the theory of (locally) compact groups which also happen to have compatible analytic atlases) there is no advantage to be gained in working over Banach spaces. The reason for this is that once the restriction 'locally compact' is put on a manifold modelled on a Banach space, then the Banach space can only be finite-dimensional. Since locally compact signifies that each point has a neighbourhood with a compact closure and since the chart maps are homeomorphisms, it follows that there must exist a point in the associated Banach space E which has a relatively compact neighbourhood. At this stage a result due to F. Riesz is applicable (see Edwards [1, p. 65]) and shows immediately that E is finite-dimensional.

Of course there is no a priori reason why we cannot drop the restriction of local compactness and study infinite-dimensional Lie groups. In fact, many results for finite-dimensional Lie groups are now known to hold in this more general setting. (See Lang [2, Chapter VI, §5], Eells [1, §3], and the citations given in this latter reference.) In spite of the attractions inherent in presenting some of these generalisations, it would mean straying too far from our chosen path to consider them here.

Local compactness aside, another point to consider before rushing willy-nilly into the realm of infinite-dimensional manifolds is that in quite general situations the manifold turns out to be nothing more than an open subset of the space on which it was modelled. To explain several of these results it is necessary to introduce an even more restrictive type of immersion than the one given in 1.2.6. An immersion $f: M \to N$ is called an embedding if it is a homeomorphism onto its image (in the induced topology). In our terminology all embeddings and immersions are analytic but clearly it makes sense to talk about C^p-embeddings and C^p-immersions. (Actually, when dealing with infinite-dimensional manifolds the definitions of immersions and embeddings must be slightly modified ... see Lang [2, p. 25 and p. 27 Proposition 2]. For the other technical terms in the results quoted below see the cited papers.)

Let E be a smooth Banach space of infinite-dimension with a Schauder basis. Suppose that M is a separable, metrizable smooth manifold modelled on E. Eells and Elworthy [1] have shown that if the tangent bundle of M is trivial, then there is a smooth embedding of M onto an open subset of E. (A manifold M modelled on a space E has a trivial tangent bundle if it is equivalent to $M \times E$. In practice this condition is not very restrictive since, for example, paracompact manifolds based on Hilbert spaces, l^p $(1 \leq p < \infty)$ and $L^p[0, 1]$ $(1 < p < \infty)$ all have such tangent bundles. These and related results are contained in Dixmier and Douady [1] and Mityagini [1].) Henderson [1] has proved a corresponding result for C^0-embeddings: suppose that M is a separable, metrizable C^0-manifold modelled on an infinite-dimensional Fréchet space F; then M admits a C^0-embedding onto an open subset of F.

Exercises

1.A. (i) Show that the two definitions of a derivative contained in formulae (1.1.1) and (1.1.2) are equivalent.

(ii) Show that if f is differentiable, then there is only one member of hom(E, F) satisfying (1.1.1).

(iii) Verify formula (1.1.3).

(iv) If f is a twice-differentiable map from an open subset U of E into F, then $f''(x) \in \hom(E, \hom(E, F)) = \hom^2(E \times E, F)$ for each x in U. Show that $f''(x)$ is symmetric, that is, that $f''(x)(u)(v) = f''(x)(v)(u)$ in F for all u, v in E.

(v) If E and F, real finite-dimensional spaces, are given bases, then each $f : E \to F$ can be expressed in the form

$$f(x_1, \ldots, x_n) = (f_1(x_1, \ldots, x_n), \ldots, f_m(x, \ldots, x_n))$$

with respect to these bases. Show that the linear map $f'(x)$ has the matrix representation

$$(\frac{\partial f_i}{\partial x_j}(x)); \qquad i = 1, \ldots, m; \qquad j = 1, \ldots, n.$$

1.B. (i) If p is a point of an analytic manifold M modelled on E and ϕ_α is a chart about p, show that given any v in E there exists ξ in $\mathcal{C}_p(M)$ such that

$$(\phi_\alpha^{-1} \xi)'(0) = v.$$

(ii) Show that the linear operations on $T_p(M)$ introduced in subsection 1.2.2 are well-defined. (For example, if $(\phi_\alpha, v) \sim_p (\phi_\beta, w)$, show that $(\phi_\alpha, \lambda v) \sim_p (\phi_\beta, \lambda w)$, and similarly for the operation of addition.)

(iii) Let f be an analytic map from $(M, (\phi_\alpha : \alpha \in A))$ into $(N, (\psi_\beta : \beta \in B))$, both analytic manifolds. Using the notation in 1.2.4, for each p in M define $f_{\#,p}$ by

$$f_{\#,p}[[\xi]]_p = [[f\xi]]_{f(p)} \cdot$$

Suppose that $f_{*,p}[\phi_\alpha, v]_p = [\psi_\beta, w]_{f(p)}$, and hence show $(\psi_\beta^{-1} f \xi)'(0) = w$ if $(\phi_\alpha^{-1}\xi)'(0) = v$, thus demonstrating that the maps $f_{\#,p}$ and $f_{*,p}$ are essentially equivalent.

1.C. **Dimension.** (i) Suppose that E and F are real finite-dimensional spaces and that M is a Hausdorff topological space which has two atlases, one modelled on E, the other on F. Show that $\dim(E) = \dim(F)$. (This proof comes down to showing that E and F are homeomorphic if and only if $\dim(E) = \dim(F)$. A proof of this latter result is given in Dugundgi [1, XVI, 6.3].)

(ii) Suppose that M is a manifold and that N is a subset of M which is also a manifold when equipped with the induced topology. Show that $\dim(M) = \dim(N)$ if and only if N is open in M. (Assume the validity of part (i) above.)

1.D. **Taylor's formula.** Let E and F be real finite-dimensional spaces and let U be an open subset of E. Suppose that $f: U \to F$ is a smooth function, that $x \in U$, $y \in E$ and that $x + ty \in U$ for all t in [0, 1]. Using $y^{(m)}$ to denote the m-tuple $(y, y, \ldots, y)$ show that

$$f(x + y) = f(x) + \frac{1}{1!} f'(x)y + \frac{1}{2!} f''(x)y^{(2)} + \ldots$$
$$\ldots + \frac{1}{(n-1)!} f^{(n-1)}(x)y^{(n-1)} + R_n,$$

where $R_n = \int_0^1 \frac{(1-t)^{n-1}}{(n-1)!} f^{(n)}(x + ty)dt(y^{(n)})$.

(Hint: Reduce the problem to the classical case by considering functions of the form $t \to \mu f(x + ty)$ on [0, 1], where μ is a linear functional on F.)

Show that R_n, when considered as a function of y, satisfies $\lim_{y\to 0} R_n(y)/\|y\|^{n-1} = 0$.

1.E. Suppose that E is a real finite-dimensional space, that X is a nonvoid set, and that for each α in some index set A, $\phi_\alpha : U_\alpha \to V_\alpha$ is a bijection from an open subset U_α of E onto a subset V_α of X. Show that if the union of the V_αs is X, and if each of the maps $\phi_\beta^{-1}\phi_\alpha$ is continuous [resp. analytic] on $\phi_\alpha^{-1}(V_\alpha \cap V_\beta)$, then

X can be given a unique Hausdorff topology such that $(X, (\phi_\alpha : \alpha \in A))$ is a manifold [resp. an analytic manifold].

1.F. Let $(M, (\phi_\alpha : \alpha \in A))$ be an analytic manifold and let p belong to V_α. A local derivation at p is a linear map $\delta : \mathfrak{F}_\alpha \rightarrow \mathbf{R}$, where $\mathfrak{F}_\alpha$ is the linear space of analytic real-valued functions on V_α, which satisfies

$$\delta(fg) = f(p)\delta(g) + g(p)\delta(f).$$

(i) Show that every element $[\phi_\alpha, v]_p$ in $T_p(M)$ gives rise to a local derivation δ at p via

$$\delta : f \mapsto \rho \circ f_{*,p}([\phi_\alpha, v]_p), \quad f \text{ in } \mathfrak{F}_\alpha,$$

where $\rho = \rho(f, p)$ is the natural projection from $T_{f(p)}(\mathbf{R})$ onto $\mathbf{R}$ given by $\rho([i, w]_{f(p)}) = w$.

(ii) Show that every local derivation at p is generated by an element in $T_p(M)$ in the above fashion.

1.G. **Product manifolds.** If M and N are analytic manifolds with analytic atlases $(\phi_\alpha : \alpha \in A)$ and $(\psi_\beta : \beta \in B)$ respectively, we define their product to be the topological space $M \times N$ equipped with the atlas $(\phi_\alpha \times \psi_\beta : \alpha \in A, \beta \in B)$. Check that this is an analytic atlas, and hence that $M \times N$ becomes an analytic manifold, and that it has the 'universal' property:

(i) the natural projections $\pi_1 : M \times N \rightarrow M$ and $\pi_2 : M \times N \rightarrow N$ are analytic; and

(ii) if P is an analytic manifold, then $f : P \rightarrow M \times N$ is analytic if and only if both $\pi_1 f$ and $\pi_2 f$ are analytic.

1.H. **Analytic continuation.** Let N, M be two analytic manifolds with M connected and let f, g be two analytic functions from M into N. If there is a nonvoid open subset of M on which f and g are equal, then f and g are equal on all of M. (Hint: This result extends the classical version of the principle of analytic continuation for nonvoid open subsets of $\mathbf{R}^n$ or $\mathbf{C}^n$; see Dieudonné [1, (9.4.2)].)

2·Lie groups and Lie algebras

The aim of this chapter is to introduce the notions of a Lie group and of its basic ancillary tools (its Lie algebra, 1-parameter subgroups, etc.) and to develop carefully their interrelationships. In the first section a Lie group is defined ... it is a topological group which can be equipped with an analytic atlas in such a way that the group operations are analytic.

After having presented a few basic examples and results on Lie groups, we introduce the fundamental tool in their study, namely the Lie algebra of a Lie group. If G is a Lie group, then the tangent space $T_e(G)$ to G at its identity e is linearly isomorphic to the space of left invariant vector fields on G which in turn is closed under the Lie product defined in the previous chapter. The Lie algebra $\underline{\underline{g}}$ of G is defined as $T_e(G)$ equipped with the Lie product induced from the space of left invariant vector fields. A further description of $\underline{\underline{g}}$ is possible in terms of the 1-parameter subgroups of G (analytic homomorphisms from R into G); each vector in $\underline{\underline{g}}$ is the tangent at e to the image of a unique 1-parameter subgroup of G and vice versa. These results lead straight into the definition of the exponential function $\exp : \underline{\underline{g}} \to G$ and the fact that $\theta : \mathbf{R} \to G$ is a 1-parameter subgroup of G if and only if it is of the form $t \mapsto \exp tX$, for some X in $\underline{\underline{g}}$.

In the last section detailed descriptions of the Lie algebras and the orthogonal group are presented. It is instructive and provides a crutch to the imagination to keep these examples in mind when reading the latter chapters.

2.1 Lie groups

We begin by reminding the reader that the definition of a topological group and their main properties are listed in Appendix A. In this regard we make the convention that the only topological groups considered are Hausdorff topological groups. Suppose that G is a topological group and let $\mu : G \times G$ and $\lambda : G \to G$ denote the corresponding product and inverse maps. An analytic atlas on G (as a topological space) is said to be compatible (with G as a group) if both μ and λ are analytic. (Refer to Exercise 1. G for the notion of a product manifold.) The following form of the definition of a Lie group, emphasising that we are primarily interested in a Lie group as a topological group, is convenient for our needs.

2.1.1 **Definition.** A topological group is said to be a Lie group if it possesses a compatible analytic atlas. The dimension of a Lie group is defined as its dimension as a manifold.

No confusion need arise as to whether or not a given compatible analytic atlas is maximal since the maximal analytic atlas generated by a compatible analytic atlas is also compatible. Moreover, it will be seen in 2.3.3 below that there is essentially only one compatible analytic atlas corresponding to each Lie group in the sense that all compatible analytic atlases generate the same maximal (compatible) analytic atlas.

2.1.2 **Examples.** (a) Let E denote an n-dimensional real or complex vector space. Then E is an additive group and as such is an n-dimensional or 2n-dimensional Abelian Lie group depending on whether it is over the real or complex field. (As mentioned in 1.1.1, the convention applying when E is a real n-dimensional vector space is that its atlas is the identity map. It is trivial that this atlas is compatible and analytic. In the case when E is over the complex field it can be written as the direct sum of F and iF, where F is a real n-dimensional space and i is a square root of -1. The atlas of E is always taken to consist of the single chart $j : F \oplus F \to E$ defined by $j(u, v) = u + iv$. Clearly this atlas is analytic and compatible.)

(b) Again suppose that E is an n-dimensional real or complex vector space and let GL(E) denote the set of all invertible (linear) endomorphisms of E. Equip GL(E) with its unique linear space topology (see A.2 of Appendix A); then GL(E) is a topological group under the operation of composition of operators and furthermore it is a Lie group. It is usually referred to as the general linear group associated with E. Since GL(E) only depends essentially on the field (real or complex) underlying E and the dimension of E, GL(n, **R**) (or GL(n, **C**)) is often written in place of GL(E). (Here **R** and **C** refer to the real and complex fields respectively.) Once a basis for E is chosen, GL(n, **R**) and GL(n, **C**) are frequently realised as the groups of $n \times n$ real and $n \times n$ complex invertible matrices respectively, with the topology of convergence of each of the entries.

We will show only that GL(n, **C**) is a Lie group. First of all define an open set U in $\mathbf{R}^{2n^2}$ by

$$(x^m_{jk} : j,\ k = 1, \ldots, n;\ m = 1,\ 2) \in U$$

if

$$(x^1_{jk} + ix^2_{jk}) \in GL(n,\ \mathbf{C})$$

and then define a homeomorphism $\phi : U \to GL(n,\ \mathbf{C})$ by $\phi(x^m_{jk}) = (x^1_{jk} + ix^2_{jk})$. (Here we have adopted the representation of GL(n, **C**) as the group of $n \times n$ complex invertible matrices so that $(x^1_{jk} + ix^2_{jk})$ signifies such a matrix with entries $x^1_{jk} + ix^2_{jk}$, $j,\ k = 1, \ldots, n$.) When GL(n, **C**) is equipped with ϕ as an atlas it becomes an analytic manifold. Let us now see that this atlas is compatible with the group structure. The product map μ is analytic provided that $\phi^{-1} \circ \mu \circ (\phi \times \phi)$ is analytic on $U \times U$. If both (x^m_{jk}) and (y^m_{jk}) belong to U, then

$$\begin{aligned}
\phi^{-1} \circ \mu \circ (\phi \times \phi)((x^m_{jk}),\ (y^m_{jk}))
&= \phi^{-1}\Big(\sum_{l=1}^{n} (x^1_{jl} + ix^2_{jl})(y^1_{lk} + iy^2_{lk})\Big) \\
&= \phi^{-1}(w^1_{jk} + iw^2_{jk}) \\
&= (w^m_{jk}),
\end{aligned}$$

where

$$w^1_{jk} = \sum_l (x^1_{jl}y^1_{lk} - x^2_{jl}y^2_{lk})$$

and

$$w_{jk}^2 = \sum_l (x_{jl}^2 y_{lk}^1 + x_{jl}^1 y_{lk}^2).$$

Thus this composite map is nothing but finite sums and products of the coordinates of the elements in U and so it is analytic. The inverse map can be shown to be analytic by a similar argument by using Cramer's rule for the inverse of a matrix. A similar argument shows that GL(n, **R**) is also a Lie group (homeomorphic to an open subset of $\mathbf{R}^{n^2}$).

(c) If E is an n-dimensional complex Hilbert space, then the subset U(E) (or U(n)) of GL(E) consisting of all the unitary endomorphisms is a compact Lie group called the unitary group of degree n. The corresponding group in the real case is denoted by O(E) (or O(n)) and is called the orthogonal group of order n.

Naturally U(E) and O(E) are equipped with their induced topologies and as such they are compact since they are closed (check this) and bounded. (Suppose that (x_{jk}) is a typical element. Then $\sum_{k=1}^{n} |x_{jk}|^2 = 1$ for each j and so the entries have moduli not exceeding 1.) It can be shown directly that they are both Lie groups, but this will not be done here since a little later it will be shown that every closed subgroup of a Lie group is itself Lie.

Theorem 6.1.1 below shows that U(E) and O(E) are to be considered as key examples of the class of compact Lie groups.

(d) Let SU(n) and SO(n) denote the subgroups of U(n) and O(n) respectively consisting of elements with determinants equal to 1. They too are compact Lie groups for the reason given in (c) above and are called the special unitary group and the special orthogonal group (of degree n) respectively.

(e) The group U(1) is more commonly known as the circle group or the 1-dimensional torus and will be denoted by T. In general, a torus is any group of the form T^n. They are compact, connected, Abelian Lie groups and it will be seen later than they are the only groups with these properties.

Extensions of local properties. To give an (algebraic) group the structure of a topological group only a certain family of sets containing the identity need be specified, the family of all left translations (or right translations) of these sets then forms a topology on G for which the group operations are continuous - thus G becomes a topological group. (See Appendix A.) Similarly, by starting with a topological group which has a chart about the identity with certain properties, then the set of all left translations (or right translations) of this chart forms a compatible analytic atlas on G. The following is typical of such results.

2.1.3 **Theorem.** Let G be a topological group and let $\phi: U \to V$ be a chart on G about its identity. If both

(i) $(u, v) \mapsto \phi^{-1}(\phi(u)\phi(v))$

(ii) $u \mapsto \phi^{-1}(\phi(u)^{-1})$

are analytic on $\{(u, v) : u, v \in U$ and $\phi(u)\phi(v) \in V\}$ and on $\phi^{-1}(V \cap V^{-1})$ respectively, then G is a Lie group.

Before proving this result we introduce the notion of the component of the identity of a topological group and prove several basic facts about it. The component of the identity of a topological group is defined as the maximal connected subset containing the identity.

2.1.4 **Lemma.** Let G_0 denote the identity component of a topological group G. Then:

(i) G_0 is a closed normal subgroup of G;

(ii) G_0 is also open whenever G is a Lie group; and

(iii) G_0 is a Lie group if and only if G is a Lie group.

Proof. (i) From the continuity of left translations, for each x in G_0, xG_o is also a connected set containing e so that $(G_0)^2 \subseteq G_0$. Similarly, the continuity of the inversion map leads to the fact that $G_0^{-1} \subseteq G_0$. Thus G_0 is a subgroup of G. If $x \in G$, then $y \mapsto xyx^{-1}$ is a continuous map whence xG_ox^{-1} is connected. But this latter set contains the identity so that it must be a subset of G_0 showing that G_0 is normal. Since G_0 is maximal as a connected set, it is closed.

(ii) Let G be a Lie group; we will prove a little more than necessary and show that G is locally connected. (A topological space is said to be locally connected if for each element x and each neighbourhood W of x the component of W to which x belongs is also a neighbourhood of x.) The whole group G is a neighbourhood of each point x in G_0 and the component of G to which x belongs is $xG_0 = G_0$. Thus if G is locally connected, G_0 must be a neighbourhood of x showing that G_0 is open.

To show that G is locally connected we first notice that, as the name implies, locally connected is a local property so that all that is needed is to show that the underlying real finite-dimensional space E of G is locally connected. This latter fact is trivial.

(iii) If G is a Lie group, then G_0 (being an open subset of G) is also a Lie group since any compatible analytic atlas on G gives rise to a compatible analytic atlas on G_0 by restricting the individual charts to G_0. On the other hand, suppose that G_0 is a Lie group and suppose that $(\phi_\alpha : \alpha \in A)$ is a compatible analytic atlas on G_0. Then the same is true of $(L_x\phi_\alpha : x \in G, \alpha \in A)$ on G_0, where L_x denotes the operator $y \mapsto xy$. This proceeds via a function similar to (2.1.3) below which relies on the analyticity of $x \to b^{-1}xb$ on G_0 for $b \in G$ (use 2.3.1). //

2.1.5. In the proof of 2.1.3 we will also need the well-known fact that each neighbourhood of the identity of a connected topological group generates the whole group. See Hewitt and Ross [1, Theorem (7.4)].

Proof (of 2.1.3). Since every neighbourhood of the identity of a topological group contains a symmetric neighbourhood (that is, a neighbourhood W satisfying $W = W^{-1}$), we can assume without loss of generality that the set V in the statement of the theorem is symmetric. Also we will assume that U, and hence V, is connected so that V lies in the identity component of G. Thus, according to Lemma 2.1.4 (iii), we can assume without loss of generality that G is connected.

Suppose that V' is an open symmetric neighbourhood of e such that $V'V' \subseteq V$. Let $U' = \phi^{-1}(V')$ and $\psi = \phi|_{U'}$; then $\psi : U' \to V'$ is a chart on G about the identity e. For each a in G define $L_a : G \to G$ by $x \mapsto ax$. Then $L_a\psi$ defines a homeomorphism from U' onto aV'

and $(L_a\psi : a \in G)$ defines an atlas on G. We first check that this atlas is analytic. Suppose that $L_a\psi$ and $L_b\psi$ are two overlapping charts; we must show that $(L_b\psi)^{-1} \circ L_a\psi$ is analytic on $(L_a\psi)^{-1}(aV' \cap bV')$, in other words, that

$$(2.1.1) \quad v \mapsto \psi^{-1}(b^{-1}a\psi(v)) \text{ is analytic on } \psi^{-1}(V' \cap a^{-1}bV').$$

Since $aV' \cap bV'$ is nonvoid, there exist v_1, v_2 in V' such that $av_1 = bv_2$. Hence $b^{-1}a = v_2v_1^{-1}$ and so $b^{-1}a \in V$. Thus there exists u in U with $\phi(u) = b^{-1}a$. Furthermore, whenever $v \in \psi^{-1}(V' \cap a^{-1}bV')$,

$$\phi(u)\phi(v) \in b^{-1}a(V' \cap a^{-1}bV') = b^{-1}V' \cap V' \subseteq V'.$$

Therefore the map in (2.1.1) can be written as

$$v \mapsto \psi^{-1}(\phi(u)\psi(v)) = \phi^{-1}(\phi(u)\phi(v))$$

and its analyticity on $\psi^{-1}(V' \cap a^{-1}bV')$ follows from 2.1.3 (i).

Before showing that the product is analytic it is convenient to show that

$$(2.1.2) \quad u \mapsto \psi^{-1}(b^{-1}\psi(u)b) \text{ is analytic at } \psi^{-1}(e)$$

for each b in G. (Theorem A.1.2 of Appendix A may be used to show that the function in (2.1.2) is well-defined on some open neighbourhood of $\psi^{-1}(e)$.) We begin by supposing that b belongs to V and hence that there exists w in U such that $\phi(w) = b$. In this case (2.1.2) is smooth since it can be written as the composition $\alpha \circ \beta$, where α and β are defined on U by

$$\alpha : u \mapsto \phi^{-1}(\phi(w')\phi(u)) \text{ with } w' = \phi^{-1}(b^{-1}),$$

$$\beta : u \mapsto \phi^{-1}(\phi(u)\phi(w))$$

both maps being analytic at $\psi^{-1}(e)$ by 2.1.3 (i). Now suppose that b belongs to G; the result stated in 2.1.5 allows b to be written as a finite product $b_1 \ldots b_k$ of elements in V. (Recall that V is symmetric and G is connected.) By writing the map in (2.1.2) as a finite composition $l_k \circ \ldots \circ l_1$ of functions

$$l_j : u \mapsto \psi^{-1}(b_j^{-1}\psi(u)b_j),$$

each of which is analytic at $\psi^{-1}(e)$ by the above argument, its own analyticity at $\psi^{-1}(e)$ is assured.

We are now in a position to show that the product map in G is analytic. From 1.1.9 the product is analytic at (a, b) provided

$$(2.1.3) \quad (u, v) \mapsto (L_{ab}\psi)^{-1}(L_a\psi(u)L_b\psi(v)) = \psi^{-1}(b^{-1}\psi(u)b\psi(v))$$

is analytic at $(\psi^{-1}(e), \psi^{-1}(e))$, since $(a, b) \in aV' \times bV'$ and $ab \in ab.V'$. This map can be expressed as the composition of

$$(u, v) \mapsto (\psi^{-1}(b^{-1}\psi(u)b), v)$$

and

$$(u, v) \mapsto \psi^{-1}(\psi(u)\psi(v)).$$

The analyticity of these two maps at $(\psi^{-1}(e), \psi^{-1}(e))$ follows from 2.1.3 (i) and (2.1.2) and consequently (2.1.3) is also analytic at this point.

The analyticity of the inverse function can be established from the analyticity of (2.1.3) and 2.1.3 (ii). //

2.2 The Lie algebra of a Lie group

In this section two further descriptions will be given of the tangent space at the identity of a Lie group G and then this space will be provided with the structure of a Lie algebra. The resulting algebra will be called the Lie algebra of G. Throughout this section it will be supposed that G represents a Lie group modelled on E.

Left invariant vector fields. If $X : G \to T(G)$ is a vector field, then it is said to be <u>left invariant</u> if

$$XL_a = (L_a)_*X \quad \text{for all } a \text{ in } G.$$

(Here L_a denotes the operator of left translation by a and is defined by $x \mapsto ax$ for all x in G.) For future applications it is useful to notice that:

$$(2.2.1)\quad (L_a)_{*,x}([\phi_\alpha, v]_x) = [L_a\phi_\alpha, v]_{L_a x}.$$

This says roughly that when the chart about $L_a x$ is chosen to be the left translation by a of a chart about x, then $(L_a)_{*,x}$ maps each vector in $T_x(G)$ to the same vector in $T_{L_a x}(G)$.

2.2.1 **Lemma.** Every left invariant vector field is analytic.

Proof. Let X denote a left invariant vector field on G. Since

$$X(x) = XL_x(e) = (L_x)_{*,e}X(e),$$

it is hardly surprising that X is analytic since its analyticity depends only on the weak analyticity of $x \mapsto (L_x)_{*,e}$. However, to prove this latter fact it is convenient to digress momentarily.

Suppose that f is an analytic function from $U \times U$ into E, where E is a real finite-dimensional space and U is a nonvoid open subset of E. If $\tilde{u}$ and $\tilde{v}$ denote the pairs (u_1, u_2) and (v_1, v_2) respectively, then whenever $\tilde{u} \in U \times U$, since f is analytic there exists an open ball $B \subseteq U \times U$ with centre $\tilde{u}$ such that for all $\tilde{u} + \tilde{v}$ in B, the series

$$\text{(i)}\qquad \sum_{m=0}^{\infty} \frac{1}{m!}\, f^{(m)}(\tilde{u})\tilde{v}^{(m)}$$

is absolutely convergent and converges to $f(\tilde{u} + \tilde{v})$.

To each u_1 in U, define a function f_{u_1} from U into E by $f_{u_1} : u_2 \mapsto f(u_1, u_2)$. The derivative $(f_{u_1})'(u_2)$ of f_{u_1} at the point u_2 is denoted by $D_2 f(u_1, u_2)$ and is referred to as the partial derivative of f with respect to the second variable. The symbol $D_1 f(u_1, u_2)$ and the notion of the partial derivative of f with respect to the first variable are defined similarly. Clearly it is possible to define higher order partial derivatives ... it follows readily that $f'(\tilde{u})(\tilde{v}) = D_1 f(\tilde{u})(v_1) + D_2 f(\tilde{u})(v_2)$ and hence that

$$\text{(ii)} \qquad f^{(m)}(\tilde{u})\tilde{v}^{(m)} = \sum_{j_1=1}^{2} \cdots \sum_{j_m=1}^{2} D_{j_1} \cdots D_{j_m} f(\tilde{u})(v_{j_1}, \ldots, v_{j_m}),$$

where the actions of the partial derivatives are commutative. (Some of the relevant details may be found in Dieudonné [1, VIII, §12] and Lang [1, V, §7].) Thus it follows that $u_1 \mapsto D_2 f(u_1, u_2)$ is analytic on U for each u_2 in U since, from (i) and (ii),

$$D_2 f(u_1 + v_1, u_2) = \sum_{m=0}^{\infty} \frac{1}{m!} D_1^m D_2 f(u_1, u_2)(v_1^{(m)}, -)$$

for u_1, u_2 in U and $(u_1 + v_1, u_2)$ in B, and the series on the right is absolutely convergent.

Returning to the Lie group G and the left invariant vector field X, we can choose a chart $\phi : W \to V$ about e which is part of a compatible analytic atlas on G. Select a nonvoid open subset U of W such that $\phi(U)$ is also a neighbourhood of e and that $\phi^{-1}(\phi(u)\phi(v)) \in W$ whenever $u, v \in U$. Suppose that $X(e) = [\phi, w]_e$; then

$$\begin{aligned} X(x) &= (L_x)_{*,e} X(e) \\ &= [\phi, (\phi^{-1} L_x \phi)'(\phi^{-1} e)(w)]_x, \end{aligned}$$

for x in $\phi(U)$. We only need to verify the analyticity of X in a neighbourhood of the identity and for this it suffices, by subsections 1.1.9 and 1.2.3, to show that $\tau^{-1} X \phi$ is analytic on U, where $\tau : (W \times E) \to \pi^{-1}(V)$ is defined by $\tau(u, v) = [\phi, v]_{\phi u}$. Now

$$\tau^{-1} X \phi(u) = (u, (\phi^{-1} L_{\phi(u)} \phi)'(\phi^{-1} e)(w))$$

for u in U, and so it is analytic on U provided

$$h : u \mapsto (\phi^{-1} L_{\phi(u)} \phi)'(\phi^{-1} e)(w)$$

is analytic on U. Define f on $U \times U$ by $f(u, v) = \phi^{-1}(\phi(u)\phi(v))$. Then f is analytic and hence

$$u \mapsto (\phi^{-1} L_{\phi(u)} \phi)'(\phi^{-1} e) = (\phi^{-1}(\phi(u)\phi))'(\phi^{-1} e)$$

$$= (f_u)'(\phi^{-1}e)$$

$$= D_2f(u, \phi^{-1}e)$$

is analytic on U following the remarks in the above paragraph. Thus h is analytic as required. //

2.2.2 Definition. A 1-parameter subgroup of a Lie group G is an analytic homomorphism θ from **R** into G.

2.2.3 Theorem. *Let G be a Lie group.*

(i) *The map $X \mapsto X(e)$ defines a one-to-one correspondence between the (analytic) left invariant vector fields X on G and the vectors of $T_e(G)$.*

(ii) *The map $\theta \mapsto \theta_{*,0}[i, 1]_0$ defines a one-to-one correspondence between the 1-parameter subgroups of G and the vectors of $T_e(G)$.*

Proof. (i) Given v in $T_e(G)$ we need to show that there exists a unique left invariant vector field X on G such that $X(e) = v$. Suppose that such a vector field exists; then

$$X(x) = XL_x(e) = (L_x)_*X(e) = (L_x)_*v$$

and so it is uniquely determined on the rest of G. On the other hand we can use this formula to define a vector field X by $X(x) = (L_x)_*v$. Clearly $X(e) = v$ and it is automatically left invariant since

$$\begin{aligned}(L_a)_*X(x) &= (L_a)_*(L_x)_*v \\ &= (L_{ax})_*v \\ &= X(ax) \\ &= XL_ax,\end{aligned}$$

for all a, x in G. It follows from 2.2.1 that X is analytic. //

Before giving the proof of part (ii) it is necessary to state a classical result in the theory of differential equations. To simplify the exposition, first we will give a more general result which will be needed later and then follow it with the particular result needed here. Before passing to the statements of these theorems, the reader should be warned that at first glance they might not look like 'classical' theorems. However, since their conclusions are only locally valid, they are translatable to results on finite-dimensional real spaces, and vice versa.

2.2.4 **Theorem.** *Let* V *be a neighbourhood of* e *in a Lie group* G *and let* W *be a neighbourhood of some point* w_0 *in* $\mathbf{R}^m$. *For each* w *in* W *suppose that* X_w *is a vector field on* G *with the further property that* $(x, w) \mapsto X_w(x)$ *is analytic on* $V \times W$. *Then there exists a neighbourhood* W_0 *of* w_0 *in* W *and an interval* $(-\varepsilon, \varepsilon)$, $\varepsilon > 0$, *in* $\mathbf{R}$ *such that for each* w *in* W_0 *there exists an analytic function* $f_w : (-\varepsilon, \varepsilon) \to G$ *such that*

$$(2.2.2) \quad (f_w)_*[i, 1]_t = X_w(f_w(t)), \quad f_w(0) = e.$$

Furthermore $(t, w) \mapsto f_w(t)$ *is analytic on* $(-\varepsilon, \varepsilon) \times W_0$, *and for fixed* w *in* W_0, *any two solutions of (2.2.2) are equal on the intersection of their domains.*

Proof. A proof can be found in most books on differential equations. Suitable references are Coddington and Levinson [1, Chapter 2, Theorem 4.1], Dieudonné [1, Theorem (10.7.1)] and Lang [2, Chapter 4, §2].

2.2.5 **Corollary.** *Suppose that* V *is an open neighbourhood of* e *in a Lie group* G *and that* X *is a vector field on* G *which is analytic on* V. *Then there exists an interval* $(-\varepsilon, \varepsilon)$, $\varepsilon > 0$, *in* $\mathbf{R}$ *and a smooth function* $f : (-\varepsilon, \varepsilon) \to G$ *such that*

$$(2.2.3) \quad f_*([i, 1]_t) = X(f(t)), \quad f(0) = e$$

on $(-\varepsilon, \varepsilon)$. *Furthermore any two solutions of (2.2.3) are equal on the intersection of their domains.*

Armed with the above corollary we can turn to the proof of part (ii) of 2.2.3 with confidence.

Proof (of 2. 2. 3 (ii)). Given a 1-parameter subgroup θ of G, then $\theta_{*,0}[i, 1]_0$ is a vector in $T_e(G)$. From part (i) of this theorem there exists a unique left invariant vector field X on G such that $X(e) = \theta_{*,0}[i, 1]_0$. Formula (2. 2. 1) combined with the facts that X is left invariant and that θ is an analytic homomorphism shows that

$$\begin{aligned}\theta_{*,t}[i, 1]_t &= \theta_{*,t}(L_t)_{*,0}[i, 1]_0\\ &= (\theta L_t)_{*,0}[i, 1]_0\\ &= (L_{\theta(t)}\theta)_{*,0}[i, 1]_0\\ &= (L_{\theta(t)})_{*,e}X(e)\\ &= X(\theta(t)),\end{aligned}$$

that is, that

$$(*) \quad \theta_{*,t}[i, 1]_t = X(\theta(t)) \quad \text{for } t \text{ in } \mathbf{R}.$$

This shows that if X is a left invariant vector field and if $\theta_{*,0}[i, 1]_0 = X(e)$ has the 1-parameter subgroup θ as a solution, then θ satisfies (*). To complete the proof of 2. 2. 3 (ii) we will show that $\theta_{*,0}[i, 1]_0 = X(e)$, and hence (*), always has a unique 1-parameter subgroup as a solution.

Given v in $T_e(G)$, from part (i) of this theorem we have the existence of a unique analytic left invariant vector field X on G such that $X(e) = v$. Applying Corollary 2. 2. 5 results in the existence of an open interval J of 0 in **R** and in analytic function $f : J \to G$ such that

$$(2.2.4) \quad f_*([i, 1]_t) = X(f(t)), \quad f(0) = e$$

for all t in J. The basic idea of the proof is to show that f is the restriction of a unique 1-parameter subgroup.

First let $J_1 \subseteq J$ be an open neighbourhood of 0 such that $J_1 + J_1 \subseteq J$. Given s in J_1 it transpires that each of the functions $f_1 : t \mapsto f(s)f(t)$ and $f_2 : t \mapsto f(s + t)$ on J_1 are solutions of (2. 2. 3) on J_1. (To see this in the case of f_1 we have

$$
\begin{aligned}
(f_1)_*([i, 1]_t) &= (L_{f(s)}f)_*([i, 1]_t) \\
&= (L_{f(s)})_* f_*([i, 1]_t) \\
&= (L_{f(s)})_* Xf(t) \\
&= XL_{f(s)}f(t) \\
&= X(f_1(t)),
\end{aligned}
$$

using (1. 2. 3), (2. 2. 4) and the left invariance of X. It is even simpler to show that f_2 is a solution.) From the uniqueness of the solution of (2. 2. 4) we have $f(s + t) = f(s)f(t)$ for all s, t in J_1. Thus f is locally a unique analytic homomorphism from $\mathbf{R}$ into G.

To show that f has a unique extension to a 1-parameter subgroup of G, let $r \in \mathbf{R}$. Then there exists a positive integer N such that $r/N \in J_1$; define $\theta(r) = (f(r/N))^N$. It is routine to verify that $\theta : \mathbf{R} \to G$ is independent of N, is an extension of f, and is a homomorphism. Also it is the unique homomorphism which extends f and is analytic since f is analytic. //

2.2.6 **Proposition.** The set of all left invariant vector fields on a Lie group is a Lie algebra when equipped with the Lie product induced from the set of all analytic vector fields on that group.

We will prove the proposition in a slightly more general form since it will then be suitable for a further application when we come to deal with subgroups of Lie groups. Let M, N denote two analytic manifolds related by an analytic function $h : M \to N$. If X and Y are analytic vector fields on M and N respectively, we say that they are h-related if $h_* \circ X = Y \circ h$. In this case, if N has the analytic atlas $(\psi_\beta : \beta \in B)$ and if $f \in \mathfrak{F}_\beta$, the set of real analytic functions on V_β, then

$$(2.2.5) \quad \tilde{Y}f \circ h = \tilde{X}(fh) \quad \text{on} \quad h^{-1}(V_\beta)$$

since
$$
\begin{aligned}
(\tilde{Y}f)(h(a)) &= f_{*,h(a)}Y(h(a)) = f_{*,h(a)}h_{*,a}X(a) \\
&= (fh)_{*,a}X(a) \\
&= \tilde{X}(fh)(a).
\end{aligned}
$$

On the other hand, if (2. 2. 5) is satisfied for all β in B and f in $\mathfrak{F}_\beta$, then the above calculations show that

$$f_{*,h(a)}Y(h(a)) = f_{*,h(a)}h_{*,a}X(a) \quad \text{for a in } h^{-1}(V_\beta),$$

for all such pairs (β, f). Arguing as in the proof of 1.3.4, we show that $Y \circ h = h_* \circ X$ and hence that X and Y are h-related. This equivalent formulation of h-relatedness can be used to show:

2.2.7 Lemma. If X_i and Y_i are h-related for $i = 1, 2$, then so are $[X_1, X_2]$ and $[Y_1, Y_2]$.

Proof. From the above paragraph,

$$(\tilde{Y}_i f) \circ h = \tilde{X}_i(fh) \quad \text{on} \quad h^{-1}(V_\beta)$$

for all functions f in $\mathfrak{F}_\beta(N)$ and $i = 1, 2$. Thus

$$\begin{aligned}([Y_1, Y_2]^{\sim}f) \circ h &= \tilde{Y}_1(\tilde{Y}_2 f) \circ h - \dots \\ &= \tilde{X}_1((\tilde{Y}_2 f)h) - \dots \\ &= \tilde{X}_1(\tilde{X}_2(fh)) - \dots \\ &= [X_1, X_2]^{\sim}(fh),\end{aligned}$$

using the result in section 1.3.3. It follows from the above discussion that $[X_1, X_2]$ and $[Y_1, Y_2]$ are h-related as required. //

Returning to the proof of Proposition 2.2.6, the notion that a vector field on a Lie group G is left invariant is precisely the same as saying that it is L_a-related to itself for all a in G. Thus if X and Y are left invariant vector fields, the above lemma shows that the same is true for $[X, Y]$. //

2.2.8 Lie algebras. Let G be a Lie group. Since the space of left invariant vector fields on G is both closed under the operation of the Lie product (1.3.2 and 2.2.6) and is canonically isomorphic to $T_e(G)$ (Theorem 2.2.3 (i)), the Lie product can be transferred to $T_e(G)$ in a natural manner. More precisely, given v, w in $T_e(G)$, we define

(2.2.6) $[v, w] = [X, Y](e)$

where X, Y are the unique left invariant vector fields on G satisfying $X(e) = v$ and $Y(e) = w$ respectively.

The Lie algebra of G is defined as $T_e(G)$ equipped with its vector space structure and the product defined in (2. 2. 6). It will be denoted by $\underline{\underline{g}}$. (If H is a Lie group, then its Lie algebra will be denoted by $\underline{\underline{h}}$.)

Because of the canonical equivalence between $\underline{\underline{g}}$ and the Lie algebra of left invariant vector fields on G, the expression '$X \in \underline{\underline{g}}$' will sometimes signify that X is a vector in $T_e(G)$ and sometimes that it is a left invariant vector field. This simplifies the notation in some of the proofs and in each case the context will save the reader from any serious confusion resulting from this notation playing a dual role.

2. 2. 9 **Exponential map.** The exponential map $\exp : \underline{\underline{g}} \to G$ is defined by setting

$$\exp(X) = \theta_X(1) \quad \text{for } X \text{ in } \underline{\underline{g}},$$

where θ_X is the 1-parameter subgroup corresponding to X as in Theorem 2. 2. 3 (ii).

From 2. 2. 3 we know that $(\theta_X)_{*,0}[i, 1]_0 = X(e)$ and this leads to the fact that the 1-parameter subgroup $\psi : t \mapsto \theta_X(st)$ satisfies $\psi_{*,0}[i, 1]_0 = sX(e)$ for all s in **R**. Thus ψ corresponds to the left invariant vector field sX and so

$$\exp(sX) = \psi(1) = \theta_X(s).$$

Hence, for s, t in R,

$$\exp((s + t)X) = \theta_X(s + t) = \theta_X(s)\theta_X(t)$$
$$= \exp(sX)\exp(tX).$$

Combining this fact with the characterisation theorem 2. 2. 3 yields:

2. 2. 10 **Theorem.** Let G be a Lie group. For each X in $\underline{\underline{g}}$, $\theta_X : t \mapsto \exp(tX)$ is a 1-parameter subgroup of G and it satisfies

$$(\theta_X)_{*,t}[i, s]_t = sX(\theta_X(t))$$

for all s, t in R. Moreover, every 1-parameter subgroup of G is of this form.

From the definition of the exponential map we know that $t \mapsto \exp(tX)$ is analytic. In Theorem 2.2.12 below we will see that exp is analytic as a function on $\underline{g}$.

2.2.11 **Inverse mapping theorem.** If M and N are analytic manifolds, and f is a map from M into N, then f is said to be a local analytic homeomorphism at p in M if:

(i) f is analytic on some open neighbourhood U of p; and

(ii) there exists an open neighbourhood V of f(p) in N and an analytic map $g : V \to M$ such that $f \circ g$ and $g \circ f$ are the identity maps on V and U respectively.

The inverse mapping theorem states that if $f : M \to N$ is analytic and has the property that $f_{*,p}$ is a linear isomorphism, then f is a local analytic homeomorphism at the point p. For a proof see Lang [1, p. 122] or almost any other book on advanced analysis.

2.2.12 **Theorem.** The exponential map is analytic and is a local analytic homeomorphism at 0.

Proof. Since the function $(x, X) \mapsto X(x)$ from $G \times \underline{g}$ into T(G) is analytic in the first variable and linear in the second, it follows that it is an analytic function. Fix X_0 in $\underline{g}$; applying Theorem 2.2.4 there exists a neighbourhood W_0 of X_0 in $\underline{g}$ and an interval $(-\varepsilon, \varepsilon)$, $\varepsilon > 0$, in **R** such that for each X in W_0 there exists an analytic function $\theta_X : (-\varepsilon, \varepsilon) \to G$ such that

$$(\theta_X)_*[i, 1]_t = X(\theta_X(t)), \quad \theta_X(0) = e$$

and furthermore that $(t, X) \mapsto \theta_X(t)$ is analytic on $(-\varepsilon, \varepsilon) \times W_0$. But Theorem 2.2.3 shows that θ_X is the restriction of a 1-parameter subgroup of G (which we shall also denote by θ_X) and so $\exp(tX) = \theta_X(t)$. The analyticity of exp in a neighbourhood of X_0 now follows from the analyticity of $(t, X) \mapsto \theta_X(t)$. Since no restrictions were placed on the selection of X_0 it follows that exp is analytic as required.

Suppose that the Lie group G is modelled on E. The map $\exp_{*,0}$ is formally a map from $T_0(\underline{g})$ into $T_e(G)$ but since both the domain and the codomain of this map are canonically isomorphic to E, then $\exp_{*,0}$ can be considered as a map from E into E. Thus, under these isomorphisms,

$$\exp_{*,0}(X) = \exp_{*,0}[i, X]_0 = [\phi_\alpha, (\phi_\alpha^{-1}\exp)'(0)(X)]_e,$$

where ϕ_α is a chart on G about e. This last expression is equal to

$$[\phi_\alpha, (\phi_\alpha^{-1}\theta_X)'(0)(1)]_e = (\theta_X)_{*,0}[i, 1]_0$$

which is none other than X(e), or X under the isomorphisms, by Theorem 2.2.3. Thus $\exp_{*,0}$ is the identity map (on E) and the inverse mapping theorem stated in 2.2.11 shows that exp is a local analytic homeomorphism. //

2.2.13 **Small subgroups.** A topological group which has a neighbourhood of the identity that doesn't contain any nontrivial subgroups is described as a group with no small subgroups. We can now prove that a Lie group has no small subgroups and in Chapter 6 we will prove the converse for compact groups.

2.2.14 **Corollary (to 2.2.12).** Every Lie group has no small subgroups.

Proof. Let G be a Lie group. From the above theorem there exists an open neighbourhood U of 0 in $\underline{g}$ on which exp is an injection. Let V be another neighbourhood of 0 which is contained in U and which satisfies $V + V \subseteq U$. Suppose also that V is selected in such a way that it is bounded. If G doesn't have the property of no small subgroups then certainly exp(V) contains a nontrivial subgroup, H say. Choose x in H with $x \neq e$. Then there exists X in V such that $\exp X = x$. We will prove that $nX \in V$ for all positive integers n, thus contradicting the boundedness of V. (Since $x \neq e$, then $X \neq 0$.)

We have $X \in V$ and suppose that $mX \in V$ for $1 \leq m < n$; then $nX = (n - 1)X + X \in V + V \subseteq U$. Also $\exp(nX) = (\exp X)^n = x^n$ and so belongs to H. But exp is one-to-one on U so that $nX \in V$ as claimed. //

2.2.15 **Remarks.** Theorems 2.2.10 and 2.2.12 contain the main positive results on the exponential map and they will be continually referred to. On the negative side, exp is neither surjective nor injective in general. (See Exercises 2.B and 2.D.) However, for large classes of groups exp is surjective. For example, it will be shown in Chapter 4 that exp has this property for all compact groups.

The condition that $\exp : \underline{g} \rightarrow G$ is surjective is of course equivalent to the condition that every element in G is contained in (the image of) a 1-parameter subgroup. In general 1-parameter subgroups are not closed, even in the case of the torus T^n, $n > 1$. (See Exercise 2.B(i).) Also when they are equipped with the induced topology they need not even be Lie groups. (See Exercise 5.B.) The importance of 1-parameter subgroups in the study of Lie groups is beyond question, but even for the general case of locally compact groups, the information gained about a particular group from a knowledge of its (continuous) 1-parameter subgroups is considerable. (See Hewitt and Ross [1, Theorem (9.1) and §25].)

2.2.16 **Direct products of Lie groups.** Let $(G_j : j \in J)$ denote a nonvoid family of topological groups; we will use $\prod_{j \in J} G_j$ to denote the Cartesian product of the G_j equipped with the product topology and the usual group structure. Then this product is also a topological group (Hewitt and Ross [1, (6.2)]) and is called the <u>direct product of the</u> G_j. Suppose further that each G_j, $j \in J$, is a nontrivial Lie group. If J is finite, then Exercise 1.G can be extended to show that their direct product is also a Lie group. If instead J is infinite, then it is readily seen that every neighbourhood of the identity in the product contains a nontrivial subgroup. Combining these facts with 2.2.14 allows us to write:

Proposition. <u>If</u> $(G_j : j \in J)$ <u>is a nonvoid family of nontrivial</u>

Lie groups, then their direct product is a Lie group if and only if J is finite.

2.3 Homomorphisms of Lie groups

If G and H are two Lie groups and f is an analytic homomorphism from G into H, then f is said to be a Lie homomorphism. If the inverse of f also exists and is analytic, then f is said to be a Lie isomorphism. The main result of this section is that every continuous homomorphism between two Lie groups is a Lie homomorphism and hence that every topological isomorphism is a Lie isomorphism.

2.3.1 **Theorem.** Every continuous homomorphism $f : G \to H$, where G and H are Lie groups, is analytic.

Proof. First of all note that it is only necessary to establish the analyticity of f in a neighbourhood of the identity. Let $\underline{\underline{h}}$ denote the Lie algebra of H. The basis of the proof is to express f in terms of analytic functions of the form $t \mapsto \exp(tY)$, where $Y \in \underline{\underline{h}}$. This will be done first when $G = \mathbf{R}$, and then in the general case.

(i) Suppose that $G = \mathbf{R}$. Let U denote a convex open neighbourhood of 0 in $\underline{\underline{h}}$ on which exp is a diffeomorphism. To any x in $\exp(\frac{1}{2}U)$ there exists a unique y in $\exp(\frac{1}{2}U)$ such that $y^2 = x$. (Such a y can be none other than $\exp(X/2)$, where $X \in \frac{1}{2}U$ and $\exp(X) = x$.) Using the continuity of f we can choose $\varepsilon > 0$ so that

$$f(t) \in \exp(\tfrac{1}{2}U) \quad \text{for } |t| \leq \varepsilon.$$

Then $f(\varepsilon) \in \exp(\frac{1}{2}U)$ and so there exists a unique X in $\frac{1}{2}U$ satisfying $f(\varepsilon) = \exp(X)$. Furthermore $f(\varepsilon/2)$ is the unique square root of $f(\varepsilon)$ in $\exp(\frac{1}{2}U)$ so that $f(\varepsilon/2) = \exp(X/2)$. Proceeding by induction yields

$$f(\varepsilon/2^n) = \exp(X/2^n) \text{ for positive integers } n,$$

and hence, for $0 < m < 2^n$,

$$f(m\varepsilon/2^n) = \exp(mX/2^n).$$

Extending this formula by continuity leads to $f(t\varepsilon) = \exp(tX)$ for all real t satisfying $|t| \leq 1$, so the analyticity of f follows from the analyticity of the 1-parameter subgroup corresponding to X in $\underline{\underline{h}}$ (Theorem 2.2.10).

(ii) We now consider the general case when G and H are arbitrary Lie groups. Let $X_1, \ldots, X_n$ be a vector basis for the Lie algebra $\underline{\underline{g}}$. Each of the maps $t \mapsto f(\exp(tX_j))$ are continuous homomorphisms from $\mathbf{R}$ into H and hence they are analytic from the above argument. These homomorphisms are thus 1-parameter subgroups of H and so, by Theorem 2.2.10, there exist elements $Y_1, \ldots, Y_n$ in $\underline{\underline{h}}$ such that

$$\exp(tY_j) = f(\exp tX_j) \quad \text{for } t \text{ in } \mathbf{R}.$$

Define a map $F : \mathbf{R}^n \to G$ by

$$F(t_1, \ldots, t_n) = \exp(t_1X_1) \ldots \exp(t_nX_n).$$

Thus $f \circ F$ is analytic since it has the form

$$f \circ F(t_1, \ldots, t_n) = \exp(t_1Y_1) \ldots \exp(t_nY_n).$$

Let e_j denote the n-tuple with 1 in the j-th place and 0s otherwise; then

$$F_{*,0}[i, e_j]_0 = \exp_{*,0}[i, X_j]_0 = [i, X_j]_0,$$

the last step following from the fact shown in the proof of Theorem 2.2.12 that $\exp_{*,0}$ is the identity map. Thus $F_{*,0}$ maps an n-dimensional basis of $\mathbf{R}^n$ into the n-dimensional basis of $\underline{\underline{g}}$ and so it is a linear isomorphism. Thus the inverse mapping theorem is applicable (see 2.2.11 above) and shows that there exists an open neighbourhood V of 0 in $\mathbf{R}^n$ on which F is a diffeomorphism. Thus, on $F(V)$, we can write $f = (f \circ F)F^{-1}$, where F^{-1} denotes the local inverse analytic function of F, showing that f is analytic in a neighbourhood of the identity as required. //

2.3.2 **Corollary.** Every topological isomorphism $f : G \to H$, where G and H are Lie groups, is a Lie isomorphism.

Proof. Apply the above result to both f and f^{-1}. //

2.3.3 **Manifold structure of Lie groups.** When we recall that the definition of a Lie group only required the existence of a compatible analytic atlas, then we see that the above results on Lie groups are open to the criticism that the atlases on the Lie groups are not specified in the statements of the theorems. Thus, for example, to talk about the Lie algebra of a Lie group G is misleading since there would be a 'Lie algebra' corresponding to each compatible analytic atlas on G. A hint to the manner in which this ambiguity can be remedied is contained in the remark immediately following Definition 2.2.1, namely that there is essentially only one compatible analytic atlas corresponding to each Lie group. We will now make this statement rigorous (and thereby show the essential uniqueness of the Lie algebra of a Lie group).

If M is a topological space equipped with analytic atlases $(\phi_\alpha : \alpha \in A)$ and $(\psi_\beta : \beta \in B)$, then these atlases are said to be equivalent if the natural embedding map $\lambda : (M, (\phi_\alpha : \alpha \in A)) \to (M, (\psi_\beta : \beta \in B))$ and its inverse are both analytic. In such a case $\lambda_{*,p}$ provides a canonical linear isomorphism between $T_p(M, (\phi_\alpha : \alpha \in A))$ and $T_p(M, (\psi_\beta : \beta \in B))$, and consequently λ_* is a bijection from $T(M, (\phi_\alpha : \alpha \in A))$ onto $T(M, (\psi_\beta : \beta \in B))$. (In fact, when the tangent bundles are equipped with the structure of analytic manifolds as explained in 1.2.3, then both λ_* and $(\lambda_*)^{-1} = (\lambda^{-1})_*$ are analytic.)

Turning to the case of Lie groups, if $(\phi_\alpha : \alpha \in A)$ and $(\psi_\beta : \beta \in B)$ are two compatible analytic atlases on a Lie group G, then the embedding map is a topological isomorphism and hence the charts are equivalent by Corollary 2.3.2 and the appropriate definitions. In particular, the corresponding Lie algebras are isomorphic as linear spaces and it is a simple step to show that they are also isomorphic as Lie algebras. (See Exercise 2.G.) Thus, up to isomorphism, there is only one Lie algebra corresponding to each Lie group.

Of course two analytic atlases are equivalent if and only if they both generate the same maximal analytic atlas. Further, the maximal analytic atlas generated by a compatible analytic atlas is also compatible and is the union of all the compatible analytic atlases. (Check these statements.) Hence a topological group has at most one maximal compatible analytic atlas.

2.3.4 **Proposition.** Let G be a Lie group and U an open neighbourhood of 0 in $\underline{g}$ on which the exponential map is an analytic homeomorphism. Then $\exp|_U$ is a member of the maximal compatible analytic atlas on G. (The chart $\exp|_U$ will be referred to as the canonical or normal chart about e in G.)

Proof. Let $(\phi_\alpha : \alpha \in A)$ be any compatible analytic chart on G. By the definition of maximal in 1.1.6 all that is required is to show that $\phi_\alpha^{-1} \circ \exp$ and $\exp^{-1} \circ \phi_\alpha$ are both analytic on their appropriate domains for each $\alpha \in A$. But this is precisely the condition that exp is an analytic homeomorphism on U. //

2.3.5. We end this section by giving an example of two analytic atlases which are not equivalent. Consider the manifolds $(\mathbf{R}, (i))$ and $(\mathbf{R}, (\phi))$, where i is the identity map on $\mathbf{R}$ and $\phi : \mathbf{R} \to \mathbf{R}$ is a non-differentiable homeomorphism. (For example, take $\phi(x) = x$ when $x < 0$ and $\phi(x) = 2x$ otherwise.) Then both the manifolds are analytic whereas this is not true for either of the embedding maps.

2.4 The general linear group.

Let $\underline{\underline{gl}}(n, \mathbf{R})$ denote the set of $n \times n$ real matrices equipped with the product

$$(2.4.1) \quad [M, N] = MN - NM,$$

and let $\underline{\underline{o}}(n)$ denote the subset of $\underline{\underline{gl}}(n, \mathbf{R})$ consisting of the skew-symmetric matrices (that is, $\underline{\underline{o}}(n) = \{M \in \underline{\underline{gl}}(n, \mathbf{R}) : M^T = -M\}$, where M^T denotes the transpose of M) also equipped with the product (2.4.1).

Evidently $\underline{\underline{gl}}(n, \mathbf{R})$ and $\underline{\underline{o}}(n)$ are Lie algebras; the main aim of this section is to establish a canonical Lie isomorphism between $\underline{\underline{gl}}(n, \mathbf{R})$ [resp. $\underline{\underline{o}}(n)$] and the Lie algebra of $GL(n, \mathbf{R})$ [resp. $O(n)$].

2.4.1 **The power series exponential map.** Let $\underline{\underline{gl}}(n, \mathbf{R})$ be equipped with the norm $M \mapsto (\sum |m_{ij}|^2)^{\frac{1}{2}}$, where the m_{ij} are the entries of M, so that it becomes a Banach space. Given M in $\underline{\underline{gl}}(n, \mathbf{R})$, it is easily verified that the matrices

$$A_m = \sum_{j=0}^{m} \frac{M^j}{j!}$$

form a Cauchy sequence in $\underline{\underline{gl}}(n, \mathbf{R})$. Thus this sequence has a limit point in $\underline{\underline{gl}}(n, \mathbf{R})$ which we denote by e^M. In other words,

$$e^M = \sum_{j=0}^{\infty} \frac{M^j}{j!} \quad .$$

(Notice that $\|e^M\| \leq \sum_{j=0}^{\infty} \frac{\|M\|^j}{j!} = e^{\|M\|}$, where $\|.\|$ is the norm on $\underline{\underline{gl}}(n, \mathbf{R})$ defined above.) Since composition in $\underline{\underline{gl}}(n, \mathbf{R})$ is continuous,

$$\begin{aligned} e^M e^{-M} &= (I + M + \frac{M^2}{2} + \dots)(I - M + \frac{M^2}{2} - \dots) \\ &= I + A(1)M + A(2)M^2 + \dots \end{aligned}$$

Whenever m is a positive nonzero integer the coefficient $A(m)$ is 0 since

$$A(m) = \sum_{j=0}^{m} \frac{1}{(m-j)!} \frac{(-1)^j}{j!} = (m!)^{-1}(1 + (-1))^m.$$

Thus $e^M e^{-M} = I$, the identity operator, and so e^M is a member of $GL(n, \mathbf{R})$. Furthermore, when M is restricted to $\underline{\underline{o}}(n)$, then $(e^M)^{-1} = e^{-M} = e^{M^T} = (e^M)^T$ (this last equality follows from Exercise 2.C(v)) and hence e^M is orthogonal. Combining the above facts yields:

2.4.2 **Proposition.** _If_ $M \in \underline{\underline{gl}}(n, \mathbf{R})$ _then_ $e^M \in GL(n, \mathbf{R})$, _and if_ $M \in \underline{\underline{o}}(n)$ _then_ $e^M \in O(n)$.

Turning to 1-parameter subgroups, each M in $\underline{\underline{gl}}(n, \mathbf{R})$ defines a homomorphism

(2.4.2) $\theta_M : \mathbf{R} \to GL(n, \mathbf{R})$ via $t \mapsto e^{tM}$.

Theorem 2.3.1 explains that for θ_M to be analytic only its continuity need be established and this only at the identity. Let $t_m \to 0$, $|t_m| \le 1$. Now

$$\|e^{t_m M} - I\| = \|t_m(M + \frac{t_m M}{2!} + \ldots)\|$$
$$\le |t_m|[\|M\| + \frac{\|M\|^2}{2!} + \ldots], \text{ since } |t_m| \le 1,$$
$$\le |t_m|(e^{\|M\|} - 1) \to 0 \text{ as } m \to \infty.$$

Thus the continuity and hence the analyticity of θ_M is assured, and hence θ_M is a 1-parameter subgroup of GL(n, **R**). But such subgroups have already been described in Theorem 2.2.10 in terms of the Lie algebra of GL(n, **R**) and the (Lie group) exponential function, and in Theorem 2.2.3 in terms of left invariant vector fields. For example, from Theorem 2.2.3 there exists a left invariant vector field X_M corresponding to each M in $\underline{\underline{gl}}(n, \mathbf{R})$ such that

(2.4.3) $(\theta_M)_{*,0}[i, 1]_0 = X_M(I)$.

Let us analyse the left hand side of this equality. This is easily done since GL(n, **R**) is (homeomorphic to) an open subset of $\mathbf{R}^{n^2}$ (see 2.1.2(b)) and so the identity chart suffices as an atlas. In this case we can write $(\theta_M)_{*,0}(s)$ as $(\theta_M)'(0)(s)$ for s in **R**. Now

$$\lim_{s\to 0} \|\theta_M(0 + s) - \theta_M(0) - sM\| . |s|^{-1}$$
$$= \lim_{s\to 0} \|(sM + \frac{(sM)^2}{2!} + \ldots) - sM\| . |s|^{-1} = 0,$$

so that, from Definition (1.1.2) of a derivative, $(\theta_M)'(0) = M$. Formula (2.4.3) now yields that $X_M(I) = M$ or, more correctly, that:

(2.4.4) $X_M(I) = [i, M]_I$ for M in $\underline{\underline{gl}}(n, \mathbf{R})$, where i denotes the identity chart on $\mathbf{R}^{n^2}$ restricted to GL(n, R). (A word of explanation may be appropriate here. Naturally GL(n, R) can be homeomorphically embedded in $\mathbf{R}^{n^2}$ in many ways. We assume that a particular embedding has been selected and that it is maintained throughout. Modulo this embedding, GL(n, R) can then be considered as an open subset of $\mathbf{R}^{n^2}$.) Following on from this, (2.4.4) shows that each M in $\underline{\underline{gl}}$(n, R) gives rise to an element $[i, M]_I$ in the Lie algebra of GL(n, R). However the dimension of GL(n, R) as a Lie group is n^2 (use Exercise 1.C(ii)) which is also the dimension of $\underline{\underline{gl}}$(n, R); thus

2.4.3 **Fact.** The correspondence $M \mapsto [i, M]_I$ is a linear bijection from $\underline{\underline{gl}}$(n, R) onto the Lie algebra of GL(n, R).

The first step in describing the Lie product in the Lie algebra of GL(n, R) is to examine more closely the left invariant vector fields X_M, $M \in \underline{\underline{gl}}(n, \mathbf{R})$, defined in (2.4.3). Since each such vector field is left invariant,

$$\begin{aligned} X_M(A) &= (L_A)_{*,I} X_M(I) \\ &= (L_A)_{*,I}[i, M]_I \quad \text{by (2.4.4)} \\ &= [i, (L_A)'(I)(M)]_A \\ &= [i, \lim_{\varepsilon \to 0} (A(I + \varepsilon M) - AI)\varepsilon^{-1}]_A \\ &= [i, AM]_A \quad \text{for } A \text{ in } GL(n, \mathbf{R}). \end{aligned}$$

Let $\rho = \rho(A)$ denote the projection $\rho : [i, B]_A \mapsto B$. From the definition in 1.3.2 of the Lie product of two vector fields X_M, X_N $(M, N \in \underline{\underline{gl}}(n, \mathbf{R}))$ at the point A in GL(n, R), we have

$$\begin{aligned} [X_M, X_N](A) &= [i, (\rho X_N)'(A)(\rho X_M(A) - (\rho X_M)'(A)(\rho X_N)(A)]_A \\ &= [i, (\rho X_N)'(A)(AM) - (\rho X_M)'(A)(AN)]_A \\ &= [i, \lim_{\varepsilon \to 0} \{(\rho X_N(A + \varepsilon AM) - \rho X_N(A))\varepsilon^{-1}\} - \ldots]_A \\ &= [i, \lim_{\varepsilon \to 0} \{((A + \varepsilon AM)N - AN)\varepsilon^{-1}\} - \ldots]_A \\ &= [i, A(MN - NM)]_A \end{aligned}$$

$$= X_{MN-NM}(A).$$

Thus $[X_M, X_N] = X_{MN-NM}$. Combining this fact with 2.4.3 and with the definition of the Lie product in the Lie algebra of a Lie group, (2.2.6) results in:

2.4.4 **Theorem.** The correspondence $M \mapsto [i, M]_I$ is a linear bijection from the Lie algebra $\underline{\underline{gl}}(n, \mathbf{R})$ onto the Lie algebra of $GL(n, \mathbf{R})$ which preserves the Lie product.

2.4.5 **Definition.** A linear map, injection or bijection from a Lie algebra which preserves the Lie products is called a Lie homomorphism, monomorphism or isomorphism, respectively.

In this terminology Theorem 2.4.4 shows that $\underline{\underline{gl}}(n, \mathbf{R})$ and the Lie algebra of $GL(n, \mathbf{R})$ are Lie isomorphic in a very simple manner. What can be said about the Lie algebra of $0(n)$?

Since $0(n)$ is a closed subgroup of $GL(n, \mathbf{R})$, a Lie group, it is a Lie group (although the proof of this general fact will not be presented until the next chapter). However the fact that $0(n)$ is a closed subgroup means that the identity chart is no longer applicable. This difficulty is surmounted in the following manner. Let $j : 0(n) \to GL(n, \mathbf{R})$ denote the identity injection; then j is continuous, hence analytic, so that Exercises 2.G and 2.H(ii) show that $j_{*,I}$ is a Lie monomorphism from the Lie algebra of G into $\underline{\underline{gl}}(n, \mathbf{R})$. (In view of Theorem 2.4.4 the Lie algebra of $GL(n, \mathbf{R})$ will be denoted by $\underline{\underline{gl}}(n, \mathbf{R})$.) By always using j and $j_{*,I}$ there is no need to deal explicitly with charts on $0(n)$.

Given M in $\underline{\underline{o}}(n)$, $t \mapsto e^{tM}$ is a 1-parameter subgroup of $0(n)$ (this follows from 2.4.2 and the argument immediately following it) and hence, from (2.4.3) and (2.4.4), there exists a unique left invariant vector field Y_M on $0(n)$ such that

$$(2.4.5) \quad Y_M(I) = (j_{*,I})^{-1}[i, M]_I.$$

Thus each M in $\underline{\underline{o}}(n)$ gives rise to the element $(j_{*,I})^{-1}[i, M]_I$ in the Lie algebra of $0(n)$. This correspondence is linear so that a dimensional argument may be used to show that all elements in the Lie algebra of

$O(n)$ are of this form.

Each element $M = (m_{ij})$ in the linear space is completely specified once the values of the entries m_{ij}, $0 < i < j \leq n$, are prescribed. (These are precisely the entries in the upper triangle of M less the diagonal entries.) There are $n(n - 1)/2$ pairs of integers (i, j) which satisfy $0 < i < j \leq n$ and so the dimension of $\underline{\underline{o}}(n)$ is $n(n - 1)/2$. The dimension of $O(n)$ as a Lie group is the dimension of $GL(n, \mathbf{R})$ as a Lie group less the number of degrees of freedom lost because of the orthogonality relations. Now $M = (m_{ij})$ is orthogonal if and only if $\sum_{k=1}^{n} m_{ij}m_{jk} = \delta_{ij}$ for $i, j \in \{1, \ldots, n\}$. (Here δ_{ij} denotes the delta function; $\delta_{ij} = 1$ if $i = j$, and 0 otherwise.) This sum is symmetric in i and j so that only the pairs (i, j) satisfying $1 \leq i \leq j \leq n$ need be considered. There are $n(n + 1)/2$ such pairs and so the dimension of $O(n)$ is $n^2 - n(n + 1)/2 = n(n - 1)/2$, the same as the dimension of $\underline{\underline{o}}(n)$.

The equality of the dimensions of (the Lie algebra of) $O(n)$ and $\underline{\underline{o}}(n)$, relation (2.4.5), the fact that $j_{*,I}$ is a Lie monomorphism and Theorem 2.4.4 combine to show:

2.4.6 Corollary. <u>The correspondence</u> $M \mapsto (j_{*,I})^{-1}([i, M]_I)$ <u>is a Lie isomorphism from</u> $\underline{\underline{o}}(n)$ <u>onto the Lie algebra of</u> $O(n)$.

The group $O(n)$ is a particularly important example of a compact Lie group since it will be seen in Chapter 6 that all compact Lie groups are closed subgroups of some $O(n)$ and hence that their Lie algebras are Lie-isomorphic to Lie subalgebras of $\underline{\underline{o}}(n)$.

Whenever $M \in \underline{\underline{gl}}(n, \mathbf{R})$, then $[i, M]_I$ is the element in the Lie algebra of $GL(n, \mathbf{R})$ corresponding to the 1-parameter subgroup θ_M of $GL(n, \mathbf{R})$ defined in (2.4.2). Thus $\exp[i, M]_I = \theta_M(1)$ and so

$$(2.4.6) \quad \exp[i, M]_I = e^M.$$

Furthermore, if G is a closed subgroup of $GL(n, \mathbf{R})$, then $j_{*,I}$ is a Lie monomorphism from $\underline{g}$ into $\underline{\underline{gl}}(n, \mathbf{R})$ (or, more precisely, into the Lie algebra of $GL(n, \mathbf{R})$) and so, for M in $j_{*,I}(\underline{g})$,

$$(2.4.7) \quad \exp \circ (j_{*,I})^{-1}(M) = e^M.$$

(This last formula is particularly easy to show when use is made of Exercise 2. F.)

Notes

Lie groups are named after the Norwegian mathematician Sophus Lie (1842-1899) following his work in the late nineteenth century on systems of differential equations. Lie was concerned with transformations of real Euclidean space, each transformation given by a differential equation, such that the composition of any two such transformations is a third such transformation which is related to the original pair in a differentiable manner. (For a translation of and a commentary on some of his work, see Lie [1].) Gradually the notion of an abstract Lie group developed and in 1935 Mayer and Thomas [1] firmly established their global study in terms of differentiable manifolds, topological groups and Lie algebras.

The concept of a 1-parameter subgroup arose from the consideration of the transformations realised after the passing of an 'infinitesimal' amount of time. This 'infinitesimal' transformation then generates a unique 1-parameter subgroup. In our terminology every (smooth) 1-parameter subgroup of a Lie group G has a unique tangent at e; this tangent belongs to the Lie algebra $\underline{g}$ of G. On the other hand, every element X in $\underline{g}$ is a tangent at the identity and so may be considered (without too serious a strain on one's imagination) as a line of 'infinitesimal' length starting from e. If the other end of this line or curve is δx, and if we suppose that movement along this line corresponds to time δt, then the curve 'generated' by δx is the image of a 1-parameter subgroup of G. Suppose further that δy corresponds to Y in $\underline{g}$ just as δx corresponds to X. Then the infinitesimal element $\delta x^{-1}.\delta y^{-1}.\delta x.\delta y$ generates a third 1-parameter subgroup of G ... the tangent at e to this subgroup is $[X, Y]$! Some justification for this is given in 3.2.9 below. (It is interesting to note that the work of Robinson on nonstandard analysis (that is, on analysis involving infinitely small and infinitely large elements) has provided a framework in which the above discussion can be made fully rigorous and meaningful. See Robinson [1, §§8.3, 8.4].) The Lie algebra of a Lie group is sometimes referred to as the infinitesimal group or the infinitesimal algebra of the group.

As was seen in the section immediately above, the Lie-group exponential function 'exp' is just an extension of the power-series exponential function 'e'. Apart from Lie groups, the function e plays an important role in the study of semigroups of operators. For example, every 1-parameter subsemigroup (with appropriate continuity conditions) of bounded operators on a Banach space B is of the form $t \mapsto e^{tM}$, where M is also a bounded operator on B. (For further details see Dunford and Schwartz [1, Chapter VIII, §1].) The finite-dimensional version of this result for subgroups is proved in Section 4 above. Note also that the very definition of exp ensures that every 1-parameter subgroup of a Lie group G is of the form $t \mapsto \exp tX$ for some X in $\underline{g}$. (See Theorem 2.2.10.)

Theorem 2.3.1 (which states that every continuous homomorphism between Lie groups is analytic) will prove to be extremely useful in the sequel. A related result has recently been proved by Kallman [1]: Let G be a compact simple Lie group; if G' is a topological group with a countable basis and $\phi : G' \to G$ is an algebraic isomorphism, then ϕ is a topological isomorphism. Time may show that this result, either combined with the previous one or on its own, will also be very useful.

Finally we remark that many authors define a Lie group to be a topological group which also possesses a smooth atlas such that the group operations are smooth. Clearly every Lie group is a smooth Lie group. In the Notes at the end of Chapter 6 references are given to classical results which show that every smooth Lie group is a Lie group. Even so, all the results of this chapter are valid if 'analytic' is replaced by 'smooth' throughout, and the proofs carry over with nothing but the occasional change of a word or a phrase. For example, every left invariant vector field on a smooth Lie group is smooth, and the exponential map from the Lie algebra of a smooth Lie group to the group itself is smooth.

Exercises

2.A. (i) Give a complete description of the tangent bundle, the vector fields, the analytic vector fields, and the left invariant vector fields of **R**.

(ii) Equip the circle group $T(= U(1))$ with a compatible analytic atlas and then repeat part (i) with **T** in place of **R**.

2.B. (i) Show that all the 1-parameter subgroups of the torus T^2 are of the form $t \mapsto (e^{iat}, e^{ibt})$, where $a, b \in \mathbf{R}$. Hence show the existence of 1-parameter subgroups of T^2 which are not closed (that is, which do not have closed images) by finding a necessary and sufficient condition for such a subgroup to be closed. If a 1-parameter subgroup of T^2 is not closed, describe its closure.

(ii) Show that in the case of **T**, exp is not an injection, but that it is surjective.

2.C. Let $M \mapsto e^M$ denote the power series exponential map defined on the Lie algebra $\underline{\underline{gl}}(n, \mathbf{R})$ as in Section 4 of this chapter. If $M, N \in \underline{\underline{gl}}(n, \mathbf{R})$ and $A \in GL(n, \mathbf{R})$, prove the following equalities:

(i) $e^{A^{-1}MA} = A^{-1}e^M A$;

(ii) If $x_1, \ldots, x_n$ are the not-necessarily-distinct eigenvalues of M, then $e^{x_1}, \ldots, e^{x_n}$ are the not-necessarily-distinct eigenvalues of e^M;

(iii) The determinant of e^M is $e^{tr(M)}$, where $tr(M)$ denotes the trace of M (the sum of its diagonal elements);

(iv) If $[M, N] = 0$, then $e^{M+N} = e^M e^N$;

(v) $e^{(M^T)} = (e^M)^T$;

(vi) $e^{(\overline{M})} = \overline{e^M}$.

2.D. **Special linear group.** Let $SL(2, \mathbf{R})$ denote the subgroup of $GL(2, \mathbf{R})$ comprised of matrices with determinant equal to 1; then $SL(2, \mathbf{R})$ is a closed subgroup of the Lie group $GL(2, \mathbf{R})$ and hence is itself a Lie group. (More specifically, if U denotes the set in $\mathbf{R}^3$

$$U = \{(x_1, x_2, x_3) : |x_1 - 1| < 1, \quad x_2 \in \mathbf{R}, \quad x_3 \in \mathbf{R}\}$$

and ϕ the map

$$\phi : (x_1, x_2, x_3) \mapsto \begin{pmatrix} x_1 & x_2 \\ x_3 & x_4 \end{pmatrix}, \quad \text{where } x_4 = (1 + x_2 x_3)x_1^{-1},$$

then ϕ is a homeomorphism between the open sets U and $\phi(U)$. Show that ϕ satisfies the conditions of Theorem 2.1.3 and hence that $SL(2, \mathbf{R})$ is a Lie group of dimension 3.)

(i) Show that the Lie algebra of $SL(2, \mathbf{R})$ is (isomorphic to) the Lie subalgebra of $\underline{\underline{gl}}(n, \mathbf{R})$ consisting of matrices with zero trace. (Hint: Use 2.C (iii).) The corresponding result for $SL(n, \mathbf{R})$ is also valid.

(ii) Show that $A_r = \begin{pmatrix} r & 0 \\ 0 & r^{-1} \end{pmatrix} \in SL(2, \mathbf{R})$ is not in the image of $\exp_{SL(2, \mathbf{R})}$ whenever $r < -1$, demonstrating that exp is not surjective in general. (Hint: Suppose that there exists M in $\underline{\underline{gl}}(2, \mathbf{R})$ such that $e^M = A_r$ and use 2.C (ii) to establish a contradiction. Of course this shows as well that $\exp_{GL(2, \mathbf{R})}$ is not surjective.)

2.E. Show that the power series exponential of $\begin{pmatrix} 0 & t \\ -t & 0 \end{pmatrix}$ is $\begin{pmatrix} \cos t & \sin t \\ -\sin t & \cos t \end{pmatrix}$, and hence find real 2×2 matrices M, N such that $e^M e^N \neq e^{(M+N)}$.

2.F. If $f : G \to H$ is an analytic homomorphism between Lie groups G and H, show that

$$f \circ \exp_G = \exp_H \circ f_{*,e}.$$

2.G. Suppose that G, H and f are as in Exercise 2.F. Show that $f_{*,e}$ is a Lie homomorphism. (Hint: First establish that $f_{*,e}(v) = w$ if and only if the left invariant vector fields corresponding to v and w respectively are f-related and then apply Lemma 2.2.7.)

Local homomorphisms. Show that the above two results remain valid in an appropriate neighbourhood of 0 in $\underline{\underline{g}}$ if f is merely a local homomorphism (see Definitions 3. 4. 3).

2. H. Suppose that G, H, and f are as in Exercise 2. G. Show that

(i) f is surjective if and only if the same is true of $f_{*,e}$ (where, for the reverse implication, we also assume that H is connected); and

(ii) f is injective implies that the same is true of $f_{*,e}$. (However the converse of this is not true in general. For example, let $\alpha \in \mathbf{R} \setminus \{0\}$ and define an analytic homomorphism $f : \mathbf{R} \to \mathbf{T}$ by $f(x) = e^{i\alpha x}$. Clearly f is not injective but describe $f_{*,0}$ and show that it is injective. Cf. Exercise 3. G.)

2. I. General affine group. (i) Let E be a (real or complex) finite-dimensional space. Denote by GA(E) the set of invertible affine transformations of E, that is, of the transformations of E into E of the form $\tau : x \mapsto Ax + a$, where $A \in GL(E)$ and $a \in E$. By writing τ in the form $\begin{pmatrix} A & a \\ 0 & 1 \end{pmatrix}$, show that GA(E) is a closed subgroup of $GL(n+1, \mathbf{R})$ or $GL(n+1, \mathbf{C})$. Hence by Theorem 3. 3. 1, it is a Lie group (of dimension $n(n + 1)$). It is known as the <u>general affine group</u> associated with E.

(ii) If $\{x, y\}$ is a basis of $\mathbf{R}^2$, define $[x, x] = 0$, $[y, y] = 0$ and $[x, y] = y$ and extend by bilinearity and antisymmetry. Show that $\mathbf{R}^2$ with this product is a Lie algebra.

(iii) Show that, up to isomorphism, the Lie algebra defined in (ii) is the unique noncommutative 2-dimensional Lie algebra. (See Exercise 3. A.)

(iv) Show that the identity component of $GA(\mathbf{R})$ is the group of transformations $x \mapsto \alpha x + \beta$ where $\alpha, \beta \in \mathbf{R}$ and $\alpha > 0$. (More generally, the identity component of $GA(n, \mathbf{R})$ is the group of transformations $x \mapsto Ax + a$, where $A \in GL(n, \mathbf{R})$ with $\det(A) > 0$ and $a \in \mathbf{R}^n$. To see this, first show that the identity component of $GL(n, \mathbf{R})$ is precisely the set of operators with positive determinants.) Denote the identity component of $GA(\mathbf{R})$ by $GA(\mathbf{R})^+$. Assuming the validity of

Exercise 3. A, show that the Lie algebra of $GA(\mathbf{R})^+$ and $GA(\mathbf{R})$ (and of all other 2-dimensional Lie groups with non-Abelian identity components) is as described in (ii).

3·The Campbell-Baker-Hausdorff formula

Our knowledge of the relationship between a Lie group and its Lie algebra is considerably strengthened by a formula due to J. E. Campbell, H. F. Baker and F. Hausdorff which was developed around the turn of this century. The Campbell-Baker-Hausdorff formula (or briefly, the CBH formula) relates the algebraic structure of the Lie algebra $\underline{g}$ of a Lie group G with the group structure of G via the exponential map. It says in part that for X, Y in some neighbourhood of 0 in $\underline{g}$, then

$$\exp(X).\exp(Y) = \exp(Y) = \exp(X + Y + \tfrac{1}{2}[X, Y] + \text{higher order terms}).$$

In this chapter we assume the version of the CBH formula applicable to abstract Lie algebras and then use it to develop the version for Lie groups. Following on from there we give two important applications of this fundamental formula. The first is that every closed subgroup of a Lie group is itself a Lie group, and the second is that if G and H are Lie groups and if there exists a Lie homomorphism F from $\underline{g}$ into $\underline{h}$, then there exists a local homomorphism f from some neighbourhood of e in G into H such that $f_{*,e} = F$. Also included is a brief discussion of simply connected Lie groups and covering groups.

3.1 The CBH formula for Lie algebras

In this section we give a fairly hurried presentation of the background necessary to meaningfully state the CBH formula for algebras. For a fuller and more rigorous treatment the interested reader is referred to Jacobson [1, Chapter 5, §5] or Magnus, Karass and Solitar [1, §5.10]. Also these works should be consulted for proofs of the result since only the statement and some motivation will be given here.

3.1.1 **Definitions.** An algebra A is defined as a real linear space equipped with an associative bilinear product and an identity. Every algebra can immediately be given the structure of a Lie algebra by defining the Lie product

$$[X, Y] = XY - YX.$$

A free algebra A_r generated by a set $\{X_j : j = 1, \ldots, r\}$ is defined as a pair (A, i) consisting of an algebra A and an injection i from $\{X_j : j = 1, \ldots, r\}$ into A such that if α is any mapping of $\{X_j : j = 1, \ldots, r\}$ into an algebra B, then there exists a unique algebra homomorphism β from A into B such that $\alpha = \beta \circ i$. (One realisation of A_r is as the linear space of formal finite $\mathbf{R}$-linear combinations of elements of the form

$$\{1\} \cup \{X_{n_1}^{e_1} \ldots X_{n_k}^{e_k} : k, e_1, \ldots, e_k \in \{1, 2, \ldots\}, \; n_j \in \{1, \ldots, r\}, n_j \neq n_{j+1}\};$$

the injection i is taken as the identity imbedding and a product is defined in the obvious manner.)

If A_r is the free algebra generated by $\{X_1, \ldots, X_r\}$, then $\overline{A}_r$ is defined as the algebra of formal power series in the X_j. That is, $\overline{A}_r$ consists of elements of the form

$$a_0 + \sum_{i=1}^{r} a_i X_i + \sum_{i,j=1}^{r} a_{ij} X_i X_j + \sum_{i,j,k=1}^{r} a_{ijk} X_i X_j X_k + \ldots ,$$

where $a_0, a_i, a_{ij}, a_{ijk}, \ldots \in \mathbf{R}$, addition and scalar multiplication are defined component by component, and if

$$b_0 + \sum_{i=1}^{r} b_i X_i + \sum_{i,j=1}^{r} b_{ij} X_i X_j + \ldots$$

is another element of $\overline{A}_r$, then these two elements are defined as being equal if $a_0 = b_0$, $a_i = b_i$, $a_{ij} = b_{ij}$, etc., and their product is defined as

$$c_0 + \Sigma_i c_i X_i + \Sigma_{ij} c_{ij} X_i X_j + \ldots ,$$

where $c_0 = a_0 b_0$, $c_i = a_0 b_i + a_i b_0$, $c_{ij} = a_0 b_{ij} + a_i b_j + a_{ij} b_0$, etc.

3.1.2 **The formal exponential map.** Given any $Z = \Sigma_i d_i X_i + \Sigma_{ij} d_{ij} X_i X_j + \ldots$ in $\overline{A}_r$, then we define e^Z as the formal series

$$(3.1.1) \quad e^Z = 1 + \frac{1}{1!} Z + \frac{1}{2!} Z^2 + \frac{1}{3!} Z^3 + \ldots .$$

Since the constant term d_0 of Z is zero, elements of degree k can only be found in the first $k + 1$ terms of (3.1.1). Thus e^Z gives rise to the element $e_0 + \Sigma_i e_i X_i + \Sigma_{ij} e_{ij} X_i X_j + \ldots$ in $\overline{A}_r$, where: $e_0 = 1$, $e_i = d_i$, $e_{ij} = d_{ij} + \frac{1}{2!} d_i d_j$, $e_{ijk} = d_{ijk} + \frac{1}{2!}(d_i d_{jk} + d_{ij} d_k) + \frac{1}{3!} d_i d_j d_k$ etc.; we will always assume that the components of e^Z have been rearranged as above so that it is an element of $\overline{A}_r$.

Let A_2 denote the free algebra generated by X and Y; then we know that both e^X and e^Y are members of $\overline{A}_2$. Furthermore

$$\begin{aligned} e^X e^Y &= (1 + X + \tfrac{1}{2!} X^2 + \ldots)(1 + Y + \tfrac{1}{2!} Y^2 + \ldots) \\ &= 1 + (X + Y) + (\tfrac{1}{2}X^2 + XY + \tfrac{1}{2}Y^2) + \ldots . \end{aligned}$$

Now let $Z = X + Y + \frac{1}{2}[X, Y]$ + higher order terms; it is apparent that

$$e^Z = 1 + (X + Y) + (\tfrac{1}{2}X^2 + XY + \tfrac{1}{2}Y^2) + \ldots$$

and so, for terms up to and including those of order 2, $e^X e^Y = e^Z$. The CBH formula affirms that we can find Z in $\overline{A}_2$, the first terms of which we already know, such that $e^X e^Y = e^Z$ and that all the components of Z are made up of higher and higher order Lie products of X and Y.

To be able to state rigorously the CBH formula the following notions are needed. Let Δ be a subset of a Lie algebra L. <u>A first order Lie product of</u> Δ is any member of Δ; an m-th order Lie product of Δ is the Lie product of any k-th order product with any (m - k)-th order Lie product, where $k \in \{1, \ldots, m - 1\}$. If Δ contains any elements of **R**, then these are termed the <u>0 order Lie products of</u> Δ. For example, use the defining relations of a Lie product to show that the second order, third order and fourth order Lie products of Δ are

$$[X, Y];$$
$$\pm[X, [Y, Z]];$$
$$\pm[W, [X, [Y, Z]]], [[W, X], [Y, Z]],$$

respectively, where $W, X, Y, Z \in \Delta$. All such Lie products are called multiple Lie products.

If A_r is the free algebra generated by $X_1, \ldots, X_r$, then an element of $\overline{A}_r$ is said to be a Lie element if it can be written in the form $\sum_{m=0}^{\infty} Z_m$ where each Z_m is a finite $\mathbf{R}$-linear combination of m-th order Lie products of $\{X_1, \ldots, X_r\}$.

3.1.3 **Campbell-Baker-Hausdorff formula.** If A_2 is the free algebra generated by elements X and Y, then there exists a Lie element Z in $\overline{A}_2$ such that Z has no term of order zero and that

$$e^X e^Y = e^Z.$$

Up to and including components of order 3, Z has the form

$$X + Y + \tfrac{1}{2}[X, Y] + \tfrac{1}{12}[[X, Y], X] + \tfrac{1}{12}[[X, Y], Y] + \ldots .$$

The validity of the first and second order terms in the expansion of Z has been established above, while that of the third order terms is left to the reader as Exercise 3.C. For a complete proof, see the references listed in the opening paragraph of this section.

3.2 The CBH formula for Lie groups

We begin with Taylor's expansion for real-valued analytic functions on charts about the identity of a Lie group. Throughout this section we suppose that G is a Lie group and that $(\phi_\alpha : \alpha \in A)$ is a compatible analytic atlas on G. Recall that if $f \in \mathfrak{F}_\alpha$, the set of real-valued analytic functions on V_α, and X is an analytic vector field on G, then $\tilde{X}f$ denotes the function

$$\tilde{X}f : x \mapsto f_{*,x}X(x), \quad x \in V_\alpha,$$

which is also a member of $\mathcal{F}_\alpha$ (see 1. 3. 3). Another standard notation to be maintained throughout this section is $\|.\|$, which will denote a norm on $\underline{g}$ describing its topology. Once selected, this norm will be supposed to be fixed throughout. Also the closed unit ball in $\underline{g}$ with centre 0 will be denoted by B.

3. 2. 1 **Lemma.** With G and $(\phi_\alpha : \alpha \in A)$ as above, let $\phi_\alpha : U_\alpha \to V_\alpha$ be a chart about e. Let V be an open neighbourhood of e in G with $V.V \subseteq V_\alpha$. Choose $\varepsilon' > 0$ so that $\exp(tB) \subseteq V$ for all $t \in I_{\varepsilon'} = (-\varepsilon', \varepsilon')$ and let X be a left invariant vector field in B. Then the function $t \mapsto f(x. \exp tX)$ is analytic on $I_{\varepsilon'}$, for each f in $\mathcal{F}_\alpha$ and x in V, and moreover there exists ε with $0 < \varepsilon \leq \varepsilon'$ such that its Taylor expansion on I_ε is

$$(3.2.1) \quad f(x. \exp tX) = \sum_{m=0}^{\infty} \frac{t^m}{m!} (\tilde{X}^m f)(x) .$$

3. 2. 2 **Remark.** If the hypotheses of the above lemma are satisfied apart from the fact that we merely require X to be a member of $\underline{g}$, then $X. \|X\|^{-1} \in B$ and we can conclude from the lemma that

$$f(x. \exp tX. \|X\|^{-1}) = \sum_{m=0}^{\infty} \frac{t^m}{m! \|X\|^m} (\tilde{X}^m f)(x)$$

for $t \in I_\varepsilon$. Hence (3. 2. 1) remains valid provided t is restricted to $(-\varepsilon\|X\|^{-1}, \varepsilon\|X\|^{-1})$.

Proof (of 3. 2. 1). Given ϕ_α and V as in the statement of the lemma, since the exponential map is continuous it is possible to choose $\varepsilon' > 0$ such that $\exp(\varepsilon' B) \subseteq V$; in particular, $\exp(tB) \subseteq V$ for all $t \in I_{\varepsilon'}$. Let $f \in \mathcal{F}_\alpha$ and $X \in B$. Since $V.V \subseteq V_\alpha$, and the function f, the product in G, and the exponential map are each analytic on their respective domains, then

$$g : t \mapsto f(x. \exp tX), \quad t \in I_{\varepsilon'},$$

is analytic. The classical version of Taylor's expansion is now applicable and allows us to write

$$g(t) = \sum_{m=0}^{\infty} \frac{t^m}{m!} g^{(m)}(0) \quad \text{for } t \in I_\varepsilon,$$

for some ε with $0 < \varepsilon \leq \varepsilon'$.

The proof is completed by showing that

(3.2.2) $g^{(m)}(0) = (\tilde{X}^m f)(x)$ for m in $\mathbf{Z}^+ = \{0, 1, 2, \ldots\}$.

To show this it is necessary to demonstrate that, on I_ε,

(3.2.3) $g^{(m)} = (\tilde{X}^m f) \circ L_x \circ \theta$ for m in $\mathbf{Z}^+$,

where θ denotes the 1-parameter subgroup $t \mapsto \exp tX$. When $m = 0$, (3.2.3) is satisfied by virtue of the definition of g; suppose that it is satisfied for the integers $0, 1, \ldots, m-1$. First note that since X is left-invariant, it is L_a-related to itself for all a in G whence, from (2.2.5), $(\tilde{X}f)L_a = \tilde{X}(fL_a)$. Thus, proceeding by induction,

(3.2.4) $(\tilde{X}^p f)L_a = \tilde{X}^p(fL_a)$ for p in $\mathbf{Z}^+$ and a in G.

Now, for s in I_ε,

$$\begin{aligned} g^{(m)}(s) &= g^{(m-1)(1)}(s) \\ &= (g^{(m-1)})_{*,s}[i, 1]_s \\ &= (\tilde{X}^{m-1}f \circ L_x \circ \theta)_{*,s}[i, 1]_s \quad \text{by (3.2.3) for } m-1, \\ &= (\tilde{X}^{m-1}(fL_x) \circ \theta)_{*,s}[i, 1]_s \quad \text{by (3.2.4)}, \\ &= (\tilde{X}^{m-1}(fL_x))_{*,\theta(s)}\theta_{*,s}[i, 1]_s \\ &= (\tilde{X}^{m-1}(fL_x))_{*,\theta(s)}X(\theta(s)) \quad \text{by 2.2.10}, \\ &= \tilde{X}(\tilde{X}^{m-1}(fL_x))(\theta(s)) \\ &= (\tilde{X}^m f) \circ L_x \circ \theta(s) \quad \text{by (3.2.4)}, \end{aligned}$$

where the penultimate line follows from the definition of $\tilde{X}$. Thus (3.2.3), and hence (3.2.2), is valid for all m in $\mathbf{Z}^+$ as required. //

3.2.3 **Corollary.** Given G and ϕ_α as in the statement of the above lemma, choose $\varepsilon > 0$ such that $\exp(tB) \subseteq V_\alpha$ for all $t \in I_\varepsilon$. For each X in B, the function $t \mapsto f(\exp tX)$ is analytic on I_ε for each f in $\mathfrak{F}_\alpha$ and its Taylor's expansion is

$$f(\exp tX) = \sum_{m=0}^{\infty} \frac{t^m}{m!} (\tilde{X}^m f)(e)$$

on some neighbourhood of 0.

3.2.4 **Remark.** If f is an analytic function on G, and X is a vector field in $\underline{g}$, denote the function $x \mapsto f_{*,x}X(x)$ on G by $\tilde{X}f$. Then $t \mapsto f(\exp tX)$ is analytic on G and the method of proof of the above lemma shows that

$$f(x.\exp tX) = \sum_m \frac{t^m}{m!} (\tilde{X}^m f)(x)$$

for x in G and t in $\mathbf{R}$.

In other words, the proof of Lemma 3.2.1 can be used to give a global version of the result. The reason that 3.2.1 is stated in terms of analytic functions on V_α and not in terms of analytic functions on the whole of G is that it is only in the first case that we know that we have sufficiently many real-valued functions to separate analytic vector fields on V_α (see the proof of 1.3.4).

3.2.5 **Subalgebras.** If L is a (real finite-dimensional) Lie algebra, and L' is a linear subspace of L which is closed under the Lie product of L, then L' is said to be a Lie subalgebra of L. Suppose that $X_1, \ldots, X_m$ are members of L; the smallest Lie subalgebra of L containing $\{X_1, \ldots, X_m\}$ is called the Lie subalgebra of L generated by $\{X_1, \ldots, X_m\}$.

The Lie subalgebra of L generated by $\{X_1, \ldots, X_m\}$ is easily seen to be the space of all finite $\mathbf{R}$-linear combinations of multiple Lie products of $\{X_1, \ldots, X_m\}$. Since this subalgebra is finite-dimensional, it also follows that it is spanned by a finite number of such multiple products.

3.2.6 **Campbell-Baker-Hausdorff formula for Lie groups.** Let G be a Lie group. There exists $\varepsilon > 0$ such that corresponding to elements X, Y in B there is a function $Z : I_\varepsilon \to \underline{g}$ such that

$$\exp tX \exp tY = \exp Z(t)$$

for $t \in I_\varepsilon$. Furthermore Z can be written as an absolutely convergent power series

$$Z(t) = \sum_{m=1}^{\infty} t^m Z_m(X, Y),$$

where $Z_m(X, Y)$ is a finite **R**-linear combination of m-th order multiple Lie products of $\{X, Y\}$. (Thus Z(t) belongs to the Lie subalgebra of $\underline{g}$ generated by $\{X, Y\}$.) Also, $Z_1(X, Y) = X + Y$ and $Z_2(X, Y) = \frac{1}{2}[X, Y]$.

3.2.7 **Remark.** A device similar to that employed in 3.2.2 shows that by suitably adjusting ε, a version of the above result can be obtained for arbitrary $X, Y \in \underline{g}$.

Proof (of 3.2.7). Suppose $X, Y \in B$ and let U be an open neighbourhood of 0 in $\underline{g}$ on which exp is a diffeomorphism. Choose $\varepsilon_1 > 0$ so that $\exp(tB)\exp(tB) \subseteq \exp(U)$ for all $t \in I_{\varepsilon_1}$. Denote the local inverse of exp on U by log and define

$$Z : I_{\varepsilon_1} \to \underline{g} \text{ by } Z(t) = \log(\exp(tX).\exp(tY));$$

clearly Z is an analytic function satisfying $Z(0) = 0$. Also

$$(3.2.5) \quad \exp(tX).\exp(tY) = \exp Z(t) \text{ for } t \in I_{\varepsilon_1}.$$

The main difficulty of the proof is to show that Z has the form described in the statement of the theorem and hence that its range is within the Lie algebra generated by $\{X, Y\}$ for t in some nonvoid I_ε. This will be done by observing that when we 'scalarise' (3.2.5) by operating on both sides by a real analytic function f defined on a suitable neighbourhood of e we obtain

$$e^{t\tilde{X}}e^{t\tilde{Y}}f(e) = e^{Z(t)^{\sim}}f(e)$$

for small t. (See (3.2.11) below.) This identity is then in a suitable form to apply the CBH formula for free algebras described in the previous section.

If only the knowledge that

$$Z(t) = t(X + Y) + \frac{t^2}{2}[X, Y] + \text{higher order terms}$$

is sought without attempting to show that $Z(t)$ lies in the Lie algebra generated by $\{X, Y\}$, the simplest path is first to write $Z(t) = tZ_1 + t^2Z_2 + t^3Z_3(t)$, where $Z_1, Z_2 \in \underline{\underline{g}}$ and Z_3 is an analytic function on some I_ε into $\underline{\underline{g}}$. Now operate on both sides of (3.2.5) by a function f as above and evaluate the first and second order terms using Lemma 3.2.1. (See, for example, Spivak [1, Chapter 10, Theorem 14] or Helgason [1, Chapter II, Lemma 1.8].)

Let $\phi_\alpha : U_\alpha \to V_\alpha$ be a chart about e and let V be an open neighbourhood of e such that $V.V \subseteq V_\alpha$. Choose $\varepsilon''_2 > 0$ such that exp(tB) is contained in V for all $t \in I_{\varepsilon''_2}$. Given f in $\mathfrak{F}_\alpha$, Lemma 3.2.1 shows that for $s, t \in I_{\varepsilon'_2}$, for some ε'_2 with $0 < \varepsilon'_2 \leq \varepsilon''_2$,

$$f(\exp sX \exp tY) = \sum_m \frac{t^m}{m!}(\tilde{Y}^m f)(\exp sX).$$

But we can also apply 3.2.1 to each of the functions $s \mapsto (\tilde{Y}^m f)(\exp sX)$ since they also belong to $\mathfrak{F}_\alpha$. Thus the analytic function $(s, t) \mapsto f(\exp sX \exp tY)$ must have the expansion

$$(3.2.6) \quad f(\exp sX \exp tY) = \sum_m \frac{t^m}{m!}\{\sum_k \frac{s^k}{k!}(\tilde{X}^k\tilde{Y}^m)f(e)\},$$

for s, t in some neighbourhood of 0 in **R**, I_{ε_2} say. Because each of the series over m and k is absolutely convergent, the series in (3.2.6) can be written as an absolutely convergent double series. If furthermore we put $s = t$, then (3.2.6) becomes

$$(3.2.7) \quad f(\exp tX \exp tY) = \sum_{k,m} \frac{t^{k+m}}{k!\,m!}\tilde{X}^k\tilde{Y}^m f(e)$$

for $t \in I_{\varepsilon_2}$.

Given t in I_{ε_2}, we can give the expression $\sum_k \frac{1}{k!}(t\tilde{X})^k$ meaning as an operator from $\mathcal{F}_\alpha$ into the set of real-valued analytic functions on V by defining

$$\{\sum_k \frac{1}{k!}(t\tilde{X})^k\}f : x \mapsto \sum_k \frac{t^k}{k!}(\tilde{X}^k f)(x)$$

for f in $\mathcal{F}_\alpha$ and x in V. (This definition is meaningful since the series on the right converges to $f(x.\exp tX)$.) Denote this operator by $e^{t\tilde{X}}$. (In effect we have shown that the formal expansion $e^{t\tilde{X}}$ can be interpreted as an actual operator on $\mathcal{F}_\alpha$ whenever $t \in I_{\varepsilon_2}$.) Define $e^{t\tilde{Y}}$ similarly. Then, by (3.2.7),

$$(3.2.8) \quad f(\exp tX \exp tY) = e^{t\tilde{X}}e^{t\tilde{Y}}f(e)$$

for f in $\mathcal{F}_\alpha$ and t in I_{ε_2}.

Returning to the function Z defined by (3.2.5), we will show that there exists $\varepsilon_3 > 0$ depending on G and B only such that

$$(3.2.9) \quad f(\exp Z(t)) = \sum_n \frac{1}{n!}\widetilde{Z(t)}^n f(e) = e^{\widetilde{Z(t)}}f(e)$$

for f in $\mathcal{F}_\alpha$ and t in I_{ε_3}. As explained in Remark 3.2.2, we can choose $\lambda > 0$ so that whenever $T \in \lambda B$, then

$$(3.2.10) \quad f(\exp T) = \sum_m \frac{1}{m!}(\tilde{T}^m f)(e).$$

(In the notation of 3.2.1 we would need $0 < \lambda < \varepsilon$.) Now select $\varepsilon_3 > 0$ with the property that

$$\log(\exp(tB).\exp(tB)) \subseteq \lambda B \text{ for } t \in I_{\varepsilon_3}.$$

This ensures in particular that $Z(t) \in \lambda B$ for $t \in I_{\varepsilon_3}$ and so (3.2.9) follows by substituting $Z(t)$ for T in (3.2.10).

In the remainder of this proof we will consider only t in I_ε, where $\varepsilon = \min(\varepsilon_1, \varepsilon_2, \varepsilon_3)$. Combining formulae (3.2.5), (3.2.8) and (3.2.9) results in

$$(3.2.11) \quad e^{t\tilde{X}}e^{t\tilde{Y}}f(e) = e^{\widetilde{Z(t)}}f(e)$$

for all t in I_ε and f in $\mathcal{F}_\alpha$.

Let A_2 denote the free algebra generated by $\{\tilde{X}, \tilde{Y}\}$. Then $e^{t\tilde{X}}e^{t\tilde{Y}} \in \bar{A}_2$ for all t and hence by the CBH formula for free algebras 3.1.3 there exists a Lie element, $Z^o(t)$ say, in $\bar{A}_2$ such that

$$(3.2.12) \quad e^{t\tilde{X}}e^{t\tilde{Y}} = e^{Z^o(t)}.$$

The element $Z^o(t)$ is an infinite series of multiple Lie products of $\tilde{X}$ and $\tilde{Y}$. (The first terms of $Z^o(t)$ are $t(\tilde{X} + \tilde{Y}) + \frac{t^2}{2}[\tilde{X}, \tilde{Y}] + \dots$.)

By virtue of (3.2.11),

$$e^{Z(t)^\sim} = e^{Z^o(t)} \quad \text{for } t \in I_\varepsilon,$$

and hence $Z(t)$ is formally an infinite power series in t, where the coefficient of t^m is a finite **R**-linear combination of m-th order Lie products of X and Y. But $t \mapsto Z(t)$ is analytic on I_ε and so the series is absolutely convergent on this domain showing that $Z(t)$ belongs to the Lie subalgebra of $\underline{g}$ generated by $\{X, Y\}$ as required. (Here, of course, $e^{t\tilde{X}}$, etc. have different interpretations in (3.2.11) and (3.2.12), the first interpretation is as an operator on $\mathcal{F}_\alpha$, the second as an element of $\bar{A}_2$. This ambiguity can be overcome by employing the canonical bijection from certain operators on $\mathcal{F}_\alpha$ (in fact, those which are infinite sums $\Sigma_i F_i$ of linear and algebraic combinations of $\{\tilde{X}, \tilde{Y}\}$ which have $\Sigma_i (F_i f)(e)$ absolutely convergent for all $f \in \mathcal{F}_\alpha$) to $\bar{A}_2$.) The fact that

$$Z(t) = t(X + Y) + \frac{t^2}{2}[X, Y] + \text{higher order terms}$$

follows from the fact that the first and second order terms of $Z^o(t)$ are $t(\tilde{X} + \tilde{Y})$ and $\frac{t^2}{2}[\tilde{X}, \tilde{Y}]$ respectively. //

3.2.8 **Order notation.** Whenever we have a map Z from some interval $I_\varepsilon = (-\varepsilon, \varepsilon)$, $\varepsilon > 0$, into a finite-dimensional space E, then this map will often be denoted by $0(t^m)$ if $Z(t).t^{-m}$ is bounded in a nonvoid neighbourhood of 0. Thus $0(t^m)$ may denote different maps at different times. With this notation, several applications of 3.2.6 yield the following:

3.2.9 **Corollary.** Given X, Y in $\underline{\underline{g}}$, then

(i) $\exp tX \exp tY \exp(-tX) = \exp(tY + t^2[X, Y] + 0(t^3))$,

(ii) $\exp(-tX)\exp(-tY)\exp(tX)\exp(tY) = \exp(t^2[X, Y] + 0(t^3))$.

3.2.10 **Remarks.** For x, y in a small neighbourhood of the identity of a Lie group G, Corollary (3.2.8)(ii) shows that

(3.2.13) $x^{-1}y^{-1}xy = \exp[\log x, \log y]$

up to terms of second order, so that the Lie product in $\underline{\underline{g}}$ provides an estimate of the 'Abelian-ness' of G. For example, a connected Lie group is Abelian if and only if $[X, Y] = 0$ for all X, Y in $\underline{\underline{g}}$. (See Exercise 3.A.) Also formula (3.2.13) should be compared with the historical discussion of the commutator of infinitesimal elements of a Lie group given in the second paragraph of the Notes at the end of Chapter 2.

3.3 Closed subgroups

The first application of the CBH formula for Lie groups will be to show that every closed subgroup of a Lie group is itself a Lie group. Application of this result is one of the more common ways of showing that a given topological group is a Lie group as, for example, was done when some standard Lie groups were being exhibited in 2.1.2.

3.3.1 **Theorem.** Let G be a Lie group and let H be a closed subgroup of G. If H is given the topology induced by G, then H is a Lie group.

Proof. The usual pattern of breaking the proof into a number of steps will be followed.

Fact 1. Let $\underline{\underline{h}} = \{X \in \underline{\underline{g}} : \exp tX \in H \text{ for all } t \text{ in } \mathbf{R}\}$; then $\underline{\underline{h}}$ is a Lie subalgebra of $\underline{\underline{g}}$.

First notice that $X \in \underline{\underline{h}}$ if and only if $tX \in \underline{\underline{h}}$ for all t in R. Let $X, Y \in \underline{\underline{h}}$ and $t \in \mathbf{R}$. From 3.2.6 for n sufficiently large,

$$\exp\frac{t}{n}X\exp\frac{t}{n}Y = \exp\{\frac{t}{n}(X+Y) + 0(n^{-2})\} \quad \text{as } n \to \infty.$$

Thus

$$(3.3.1) \qquad \{\exp\frac{t}{n}X\exp\frac{t}{n}Y\}^n = \exp(t(X+Y) + 0(\frac{1}{n})) \quad \text{as } n \to \infty.$$

Since $X, Y \in \underline{\underline{h}}$, then the same is true of $\frac{t}{n}X$ and $\frac{t}{n}Y$ so that the left hand side of (3.3.1) always belongs to H. However H is closed and so

$$\lim_{n\to\infty}(\exp\frac{t}{n}X\exp\frac{t}{n}Y)^n = \exp(t(X+Y)) \in H.$$

Thus from the definition of $\underline{\underline{h}}$, $X + Y \in \underline{\underline{h}}$ and so $\underline{\underline{h}}$ is a linear space.

The fact that $\underline{\underline{h}}$ is closed under the Lie product in $\underline{\underline{g}}$ follows by a similar argument to that given in the above paragraph by deducing

$$\exp t^2[X, Y] = \lim\{\exp(-\frac{t}{n}X)\exp(-\frac{t}{n}Y)\exp(\frac{t}{n}X)\exp(\frac{t}{n}Y)\}^{n^2}$$

from Corollary 3.2.8 (ii).

<u>Fact 2.</u> If we split $\underline{\underline{g}}$ as $L \oplus \underline{\underline{h}}$, where L is a subspace of $\underline{\underline{g}}$, then there exists a neighbourhood W of 0 in L such that given $X \in W$, $X \neq 0$, then $\exp X \notin H$.

Suppose the contrary, that there exists a sequence (X_m) in L with $\lim_m X_m = 0$ such that $\exp X_m \in H$ for all m. Equip L with a Euclidean norm $\|.\|$. By passing to a subsequence if necessary, the sequence $(X_m/\|X_m\|)$ has a limit point, X say, and this limit point belongs to L. Furthermore, $\|X\| = 1$ and so $X \neq 0$. The required contradiction comes from demonstrating that $X \in \underline{\underline{h}}$. Given t in **R**, since $\|X_m\| \to 0$ as $m \to \infty$, then a sequence (n_m) of integers can be chosen so that $\lim_m n_m.\|X_m\| = t$. Thus

$$\exp tX = \lim_m \exp tX_m/\|X_m\| = \lim_m \exp n_m.X_m$$
$$= \lim_m(\exp X_m)^{n_m},$$

showing that $\exp tX \in H$, since H is closed. Hence $X \in \underline{\underline{h}}$.

<u>Fact 3.</u> There exist neighbourhoods U of 0 in $\underline{\underline{h}}$ and V of

e in G such that

$$\exp U = V \cap \exp \underline{\underline{h}} = V \cap H.$$

Considering a direct sum $\underline{\underline{g}} = \underline{\underline{h}} \oplus L$ as in Fact 2, neighbourhoods U of 0 in $\underline{\underline{h}}$ and U' of 0 in L can be chosen such that

$$\phi : \underline{\underline{g}} \to G \text{ defined by } \phi(X) = \exp(X_1)\exp(X_2)$$

is a diffeomorphism on $U \oplus U'$. (Here X is decomposed as $X_1 + X_2$.) Let us also take $U' \subseteq W$, where W is as described in Fact 2. Define $V = \phi(U \oplus U')$. Given x in $V \cap H$, then $x = \exp X_1 \exp X_2 \in H$, where $X_1 \in U$ and $X_2 \in U'$, showing that $\exp X_2 \in H$. Thus from Fact 2, $X_2 = 0$ and hence $x \in \exp U$. Thus

$$V \cap \exp \underline{\underline{h}} \subseteq V \cap H \subseteq \exp U.$$

On the other hand, given $X \in U \subseteq \underline{\underline{h}}$, then

$$\exp X = \exp X \exp 0 = \phi(X) \in V \cap \exp \underline{\underline{h}},$$

completing the proof of Fact 3.

Fact 4. There exists an open neighbourhood U_1 of 0 in $\underline{\underline{h}}$ such that $\exp|_{U_1}$ satisfies the conditions of Theorem 2.1.3.

Let U be as in Fact 3. From the CBH formula for Lie groups (Theorem 3.2.6), Fact 1 above and the fact that U is a neighbourhood of 0 in $\underline{\underline{h}}$, we can choose an open neighbourhood of 0 in U such that $\log(\exp X \exp Y) \in \underline{\underline{h}}$ for X, Y in this neighbourhood. (Here we have already assumed that we are working inside some neighbourhood of 0 in $\underline{\underline{g}}$ on which exp is a diffeomorphism.) Combining this with Proposition 2.3.4 shows that we can choose an open neighbourhood U_1 of 0 in U such that $\exp|_{U_1}$ is a homeomorphism and

$$(3.3.2) \quad (X, Y) \mapsto \log(\exp X \exp Y)$$

is an analytic map from $\{(X, Y) : X, Y \in U_1 \text{ and } \exp X . \exp Y \in \exp U_1\}$ into U_1. Also

(3.3.3) $X \mapsto \log((\exp X)^{-1}) = -X$

is clearly analytic on $\{X : X \in U_1 \text{ and } (\exp X)^{-1} \in U_1\}$. Formulae (3.3.2) and (3.3.3) show that the conditions of Theorem 2.1.3 are satisfied by $\exp|_{U_1}$.

This completes the proof of Theorem 3.3.1 since it follows from Theorem 2.1.3 that H is a Lie group. //

Remark. Long before getting to this point the reader will have surmised that the Lie algebra of H is (Lie isomorphic to) the Lie algebra $\underline{\underline{h}}$ introduced in the above proof. This surmisal will be verified in Chapter 5.

3.4 Simply connected Lie groups

Roughly speaking a topological space is said to be simply connected if it has no 'holes'. The main result on simply connected topological groups is that if G is a simply connected topological group and H is an (algebraic) group, then any map f from an open connected neighbourhood U of e in G into H which satisfies $f(xy) = f(x).f(y)$ whenever x, y and xy lie in U can be extended to a homomorphism of G into H. We will not prove this result here (for a proof see Hausner and Schwartz [1, p. 39] or Hochschild [1, p. 54]) but will use it to show that if G and H are Lie groups, where G is also simply connected, and if $F : \underline{\underline{g}} \to \underline{\underline{h}}$ is a Lie homomorphism, then there exists an analytic homomorphism $f : G \to H$ such that $f_{*,e} = F$.

This result is a converse for simply connected Lie groups of the result contained in Exercise 2.G.

3.4.1 **Definitions.** Let M be a Hausdorff topological space.

(i) A <u>path</u> in M is a continuous map $f : [0, 1] \to M$. Its <u>end points</u> are $f(0)$ and $f(1)$, and if $f(0) = f(1)$ then the path is said to be <u>closed</u>.

(ii) The space M is said to be <u>path connected</u> if every pair of points in M are the end points of some path.

(iii) The space M is said to be <u>simply connected</u> if it is path connected and if there exists a point p_o in M such that corresponding to every closed path f in M with end point p_o there exists a continuous map $h : [0, 1] \times [0, 1] \to M$ such that

$$h(s, 0) = f(s) \text{ and } h(s, 1) = p_o \text{ for all } s \text{ in } [0, 1]$$
$$h(0, t) = p_o = h(1, t) \text{ for all } t \text{ in } [0, 1].$$

3. 4. 2 **Discussion.** It is clear that every path connected space is connected and Exercise 3. D (ii) shows that every connected manifold is path connected. Thus a manifold is connected if and only if it is path connected.

The definition of simply connected for a path connected space can be thought of as the requirement that there exists a fixed point such that every closed path through this point can be continuously deformed to this point (in other words, the space has no 'holes'). The study of paths and their deformations is called homotopy theory. Briefly two paths are said to be equivalent if one can be continuously deformed into the other. Also an obvious group structure can be put on the family of closed paths through some point, the product of two paths being the path obtained by following the first path, then the second. The group obtained modulo the above equivalence relation is called the fundamental group of the space. In this terminology, which covers the basic notions of homotopy theory, a space is seen to be simply connected if and only if it is path connected and its fundamental group is trivial.

3. 4. 3 **Definitions.** Let G and H be Lie groups. If f is an analytic map on some open neighbourhood U of e in G into H which satisfies $f(x).f(y) = f(xy)$ for all x, y and xy in U, then f is called a local homomorphism (for Lie groups). If in addition f^{-1} exists as an analytic function on some neighbourhood of e in H, then f is said to be a local isomorphism, in which case G and H are said to be locally isomorphic.

Exercise 2. G required a proof of the fact that if G and H are Lie groups and if $f : G \to H$ is an analytic homomorphism, then $f_{*, e} : \underline{\underline{g}} \to \underline{\underline{h}}$ is a Lie homomorphism. An obvious modification of this proof shows that the result remains valid when f is merely a local homomorphism.

3.4.4 **Theorem.** Let G and H be Lie groups, and let $F : \underline{\underline{g}} \to \underline{\underline{h}}$ be a Lie homomorphism. Then there exists a local homomorphism f from a neighbourhood of e in G into H such that $f_{*,e} = F$. This local homomorphism is unique in the sense that if there exists a second local homomorphism with this property, then it must be equal to f on some neighbourhood of e in G.

Proof. If there exists a local homomorphism f from some neighbourhood of e in G into H satisfying $f_{*,e} = F$, then the second part of Exercise 2.F shows that $f \circ \exp_G = \exp_H \circ F$ in a neighbourhood of 0 in $\underline{\underline{g}}$. Thus the uniqueness of the local homomorphism is established, and we will use the above formula to show the existence of a suitable local homomorphism.

Let U_G be a neighbourhood of 0 in $\underline{\underline{g}}$ on which exp (that is, $\exp_G$) is a diffeomorphism, and let V_G be an open neighbourhood of 0 in U_G such that the function

$$\eta_G : (X, Y) \mapsto \log(\exp X \exp Y)$$

is analytic on $V_G \times V_G$ and on this domain can be expressed as an infinite sum $\Sigma_m r_m Z_m(X, Y)$, where the r_m are real numbers and each $Z_m(X, Y)$ is a multiple Lie product of elements in $\{X, Y\}$. (See 3.2.6.) Choose U_H, V_H and η_H similarly for the group H. Also choose U_G with the further requirement that $F(U_G) \subseteq U_H$. (This is possible since F is linear and hence continuous.) Thus for X, Y in V_G,

$$F \circ \eta_G(X, Y) = F(\Sigma_m r_m Z_m(X, Y))$$
$$= \Sigma_m r_m Z_m(F(X), F(Y)) = \eta_H(F(X), F(Y)),$$

since F is a Lie homomorphism.

Define $f : \exp(V_G) \to H$ by $f(\exp X) = \exp \circ F(X)$ for X in V_G. If x, y and xy lie in $\exp(V_G)$, then there exists X, Y in V_G such that $\exp X = x$, $\exp Y = y$, $\exp \eta_G(X, Y) = xy$, and $\eta_G(X, Y) \in V_G$. Thus the properties of η_G, η_H and F established above and the definition of f together yield

$$\begin{aligned} f(xy) &= f(\exp(\eta_G(X, Y))) = \exp F \circ \eta_G(X, Y) \\ &= \exp(\eta_H(F(X), F(Y))) = \exp F(X). \exp F(Y) \\ &= f(\exp X). f(\exp Y) = f(x). f(y), \end{aligned}$$

showing that f is a local homomorphism as required. //

The following two corollaries of Theorem 3.4.4 are obtained by using the extension result stated in the opening paragraph of this section (§3.4).

3.4.5 **Corollary.** Let G and H be Lie groups, where G is simply connected. If $F : \underline{g} \to \underline{h}$ is a Lie homomorphism, then there exists a unique analytic homomorphism $f : G \to H$ such that $f_{*,e} = F$.

3.4.6 **Corollary.** Let G and H be simply connected Lie groups. Then G and H are isomorphic as Lie groups if and only if $\underline{g}$ and $\underline{h}$ are isomorphic as Lie algebras.

3.4.7 **Covering groups.** The theory of (simply connected) covering groups is of vital importance to the local study of Lie groups. The crux of the main result in this theory is that corresponding to each Lie group G there exists a simply connected Lie group $\tilde{G}$ such that $\tilde{G}$ and G are locally isomorphic. (For a proof of this beautiful result see Hausner and Schwartz [1, p. 42] or Hochschild [1, p. 134].) From Corollary 3.4.6 we know that $\tilde{G}$ is the only simply connected Lie group with this property - $\tilde{G}$ is called the covering group of G. The importance of the notion of locally isomorphic Lie groups follows from the fact that two Lie groups are locally isomorphic if and only if their Lie algebras are Lie-isomorphic (Exercise 3.E). Moreover we have just seen that in every family of locally isomorphic Lie groups there exists precisely one simply connected Lie group and this group is the covering group of each of the groups in the family.

If G is a Lie group with covering group $\tilde{G}$, let $p : \tilde{G} \to G$ denote a homomorphism from $\tilde{G}$ into G which extends the local isomorphism between $\tilde{G}$ and G. The expression '$p : \tilde{G} \to G$ is the simply-connected covering group of G' will signify that p, $\tilde{G}$ and G are related as in the preceding sentence. Since p is a local diffeomorphism, $p(\tilde{G})$ is a neighbourhood of the identity in G and so 2.1.5 can be used to show that p is surjective whenever G is connected.

Notes

The formula which we have exhibited under the names of Campbell, Baker and Hausdorff ... that is, the CBH formula ... is sometimes referred to as the Baker-Hausdorff formula (Magnus, Karass and Solitar [1]), the Campbell-Hausdorff (Hochschild [1]) or, as we have done, the Campbell-Baker-Hausdorff formula (Hausner and Schwartz [1]). The formula was first given in 1898 in terms of exponentials of (linear) operators (of function spaces). See Campbell [1]. However no consideration was given to the question of convergence, as was necessary in the statement of the formula for Lie groups given above. This gap was remedied (or rather, avoided) independently by Baker in 1905 (see Baker [1]) and Hausdorff in 1906 (see Hausdorff [1]) by developing the algebraic version of the CBH formula given in 3.1.3 for noncommutative algebras over **R**. The role of the CBH formula is to link the group structure of a Lie group with the algebraic structure of its Lie algebra via the exponential map. That it is of fundamental importance can be seen from the two applications given above.

Some versions of the CBH formula for Lie groups state only that if G is a Lie group, then

$$\exp tX \exp tY = \exp\{t(X + Y) + \frac{t^2}{2}[X, Y] + 0(t^3)\}$$

for X, Y in $\underline{g}$. Others go further and state also that there exists $\varepsilon > 0$, $\varepsilon = \varepsilon(X, Y)$ such that $0(t^3)$ belongs to the Lie subalgebra of $\underline{g}$ generated by $\{X, Y\}$ provided $|t| < \varepsilon$. Here it seemed convenient to establish the two applications of the CBH formula by using its full strength even though this is not strictly necessary (see Helgason [1]).

Thus in the version of the CBH formula given above, care was taken to choose $\varepsilon > 0$ first, so that whenever $X, Y \in B$ and $|t| < \varepsilon$, then $0(t^3)$ belongs to the Lie subalgebra of $\underline{g}$ generated by $\{X, Y\}$. If it is only known that $X, Y \in \underline{g}$, then the procedure used in 3.2.2 shows that the CBH formula remains valid for

$$|t| < \varepsilon . \min\{ \|X\|^{-1}, \|Y\|^{-1} \}.$$

Exercises

3.A. A Lie algebra is said to be commutative if $[X, Y] = 0$ for all X, Y in L. Show that a connected Lie group is Abelian if and only if its Lie algebra is commutative.

3.B. Let G be an Abelian Lie group with Lie algebra $\underline{g}$.

(i) Considering only the additive structure of $\underline{g}$, show that $\exp : \underline{g} \to G$ is a group homomorphism.

(ii) Show that the kernel of the exponential map is discrete and that it is of the form $ZX_1 + \ldots + ZX_m$, where Z denotes the group of integers and $\{X_1, \ldots, X_m\}$ is a linearly independent set of vectors in $\underline{g}$.

(iii) Show that $\exp : \underline{g} \to G$ is surjective when G is connected (and Abelian) and hence that G is a connected Abelian Lie group if and only if it is isomorphic to a group of the form $T^m \times R^{n-m}$, where n is the dimension of G. (This result is shown for compact groups in Chapter 6.)

(iv) Describe all the compact Abelian Lie groups.

3.C. Verify the third order terms for Z in the CBH formula 3.1.3.

3.D. (i) Show that every path connected Hausdorff topological space is connected.

(ii) Show that every connected manifold is path connected. (Hint: Given p in a connected manifold M, let M_p denote the subset of M consisting of points which can be connected to p by paths and show that M_p is both open and closed.) See also Remark 4.1.3 (ii).

3. E. Use Theorem 3. 4. 4 to show that two Lie groups are locally isomorphic if and only if their Lie algebras are Lie-isomorphic.

3. F. Describe fully all the 1-dimensional Lie algebras and hence show that **T** and **R** are the only connected 1-dimensional Lie groups.

3. G. Let $p : \tilde{G} \to G$ be a simply connected covering group of a connected group G. Show that p is not injective whenever G is not simply connected. (Since $p_{*,e}$ is always injective, this result is an extension of Exercise 2. H (ii).)

3. H. Let G be a Lie group and let $X, Y \in \underline{g}$. Define the associated 1-parameter subgroups θ_1 and θ_2 by

$$\theta_1(t) = \exp(tX), \quad \theta_2(t) = \exp(tY)$$

for $t \in \mathbf{R}$. Prove that the limits

$$\theta_3(t) = \lim_{n\to\infty} \{\theta_1(t/n)\theta_2(t/n)\}^n$$
$$\theta_4(t) = \lim_{n\to\infty} \{\theta_1(-t^{\frac{1}{2}}/n)\theta_2(-t^{\frac{1}{2}}/n)\theta_1(t^{\frac{1}{2}}/n)\theta_2(t^{\frac{1}{2}}/n)\}^{n^2}$$

exist for all $t \in \mathbf{R}$ by showing that θ_3 and θ_4 are the 1-parameter subgroups

$$\theta_3(t) = \exp(t(X + Y)), \quad \theta_4(t) = \exp(t[X, Y]).$$

4 · The geometry of Lie groups

The two main results in this chapter concern a class of Lie groups which contains all compact connected Lie groups, namely the class of complete connected Lie groups which possess invariant Riemannian metrics. For groups in this class it is first shown that geodesics and translates of 1-parameter subgroups are essentially the same (Theorem 4. 3. 3) and then, as a corollary, that the associated exponential map is surjective (Corollary 4. 3. 5). These results are presented solely for their own interest and are not needed to establish the main structure Theorems described in Chapter 6.

The first section of this chapter contains a brief discussion of Riemannian manifolds (with some statements being given without proof) while the second section is concerned primarily with the establishment of a necessary and sufficient condition for a Lie group to possess an invariant Riemannian metric. In the third section these preliminaries culminate with the proofs of the main results described above.

4. 1 Riemannian manifolds

Throughout this section we will suppose that M is an analytic manifold. By selecting p in M and imposing a norm on $T_p(M)$ we can arrive at a notion of distance in an 'infinitesimal' neighbourhood of p in the following way. Temporarily ignoring all the usual caveats against talking about infinitesimals, let q be a point in this neighbourhood of p and let ξ be an analytic curve on M with $\xi(0) = p$ and $\xi(\delta t) = q$. The distance between p and q is defined as δt times the norm of the tangent of ξ at p. Furthermore, if a norm is placed on each of the spaces $T_p(M)$, $p \in M$, then we have an approach to the notion of the length of a curve in M by dividing the curve into 'infinitesimal' parts, determining the length of each part, and then summing (that is, integrating) these lengths. The details necessary to formalise the above discussion will be presented in this section.

4.1.1 **Definition.** An analytic manifold M is said to be Riemannian, or to have a Riemannian metric, if corresponding to each p in M there exists an inner product (that is, a positive definite, symmetric, bilinear form) $\langle , \rangle_p$ on $T_p(M)$ such that the map

$$p \mapsto \langle X(p), Y(p) \rangle_p \quad \text{from } M \text{ into } \mathbf{R}$$

is analytic for each pair of analytic vector fields X, Y.

Once we have a Riemannian metric it is a simple matter to introduce a notion of length for certain curves. Let a, b be points in **R** with $a \leq b$. If ξ is an analytic map from $[a, b]$ into M (that is, if ξ can be extended to an analytic map from $(a - \varepsilon, b + \varepsilon)$ into M for some $\varepsilon > 0$), then ξ is said to be an analytic curve in M, or simply a curve in M. A broken analytic curve in M is a map $\xi : [a, b] \to M$ such that for some partition of $[a, b]$, ξ is analytic on each of the closed subintervals.

By an obvious modification of these definitions we allow the possibility that $a = -\infty$, $b = \infty$, or both. In the case where both a and b are finite, then the endpoints of ξ are defined as $\xi(a)$ and $\xi(b)$.

4.1.2 **Definition.** If $[a, b]$ is a bounded interval in **R**, and if $\xi : [a, b] \to M$ is an analytic curve in a Riemannian manifold M, then the length of ξ is defined as

$$L_{p,q}(\xi) = \int_a^b \| \xi_{*,t}(1) \|_{\xi(t)} dt,$$

where p, q are the endpoints of ξ and $\| \xi_{*,t}(1) \|^2_{\xi(t)} = \langle \xi_{*,t}(1), \xi_{*,t}(1) \rangle_{\xi(t)}$. Similarly if ξ is broken analytic, then its length is defined as the sum of the lengths of each of the analytic parts.

4.1.3 **Remarks.** (i) Clearly the length of a curve depends upon the particular Riemannian structure defined on the manifold - different Riemannian metrics in general leading to different lengths of the same curve. On the other hand, the length of a curve is independent of the parameters used to describe it. (See Exercise 4. B.)

(ii) In Exercise 3. D it was stated that a manifold is connected if and only if it is path connected. The method suggested to prove this can also be used to show that in a connected manifold every pair of points can be joined by a broken analytic curve.

4.1.4 **Example.** In the case of the analytic manifold $\mathbf{R}^n$, $T_p(\mathbf{R}^n) = \mathbf{R}^n$ for each p in $\mathbf{R}^n$. Given x, y in $\mathbf{R}^n$, define

$$\langle x, y \rangle_p = \sum_{i=1}^{n} x_i y_i ,$$

where $x = (x_1, \ldots, x_n)$ and $y = (y_1, \ldots, y_n)$. In this way we have a Riemannian metric on $\mathbf{R}^n$ since $\langle\ ,\ \rangle_p$ is an inner product for each p and $p \mapsto \langle X(p), Y(p) \rangle_p$ is analytic from $\mathbf{R}^n$ into $\mathbf{R}$ for each pair of analytic vector fields X, Y. (See Exercise 4. A for further details.)

If $\xi : [a, b] \to \mathbf{R}^n$ is an analytic curve, then it can be written in terms of coordinates on $\mathbf{R}^n$ as $\xi = (\xi_1, \ldots, \xi_n)$, where $\xi(t) = (\xi_1(t), \ldots, \xi_n(t))$. In terms of the above Riemannian metric, the length of this curve is

$$\int_a^b \| \xi_{*,t}(1) \| \,dt = \int_a^b (\langle \xi_{*,t}(1), \xi_{*,t}(1) \rangle)^{\frac{1}{2}} dt$$
$$= \int_a^b (\xi_1'(t)^2 + \ldots + \xi_n'(t)^2)^{\frac{1}{2}} dt,$$

the last term being the usual expression for the length of a curve in $\mathbf{R}^n$.

4.1.5 **Distance.** Given a connected Riemannian manifold M, the <u>distance</u> d(p, q) between two points p, q in M is defined as

$$d(p, q) = \inf \{ L_{p,q}(\xi) : \xi \text{ is a broken analytic curve with endpoints } p \text{ and } q \}.$$

4.1.6 **Metrics.** A <u>metric space</u> is a pair (X, l) where X is a set and l is a function from the Cartesian product $X \times X$ into the non-negative reals such that for all points x, y and z of X:

(i) $l(x, y) = l(y, x)$;

(ii) $l(x, z) \le l(x, y) + l(y, z)$;

(iii) $l(x, x) = 0$; and

(iv) if $l(x, y) = 0$, then $x = y$.

Where no confusion is likely, a metric space (X, l) is often referred to simply as X. If (X, l) is a metric space, then it will always be thought of as a Hausdorff topological space, a subbase for the topology on X being given by the open spheres $\{y : l(x, y) < r\}$, where r ranges over all the positive numbers and x ranges over X. This topology is referred to as the metric topology for (X, l).

A sequence (x_n) in a metric space is said to be Cauchy if for each $\varepsilon > 0$ there exists an integer N such that $l(x_n, x_m) < \varepsilon$ for $m, n > N$. If every Cauchy sequence has a limit point, then the space is said to be complete. For further details in the theory of metric spaces, see Kelley [1, Chapter 4].

The following result shows that the topology of a Riemannian manifold is a metric topology, and hence the terminology 'Riemannian metric'.

4.1.7 **Theorem.** Let M be a Riemannian manifold. Then the corresponding distance function is a metric and moreover, the metric topology derived from the distance function is equivalent to the original topology on M.

Proof. Let $(M, (\phi_\alpha : \alpha \in A))$ denote a Riemannian manifold with Riemannian metric $p \mapsto \langle\,,\,\rangle_p$ and let d denote the associated distance function. It is trivial that d satisfies the first three requirements of 4.1.6 for it to be a metric. Select $q \in M$ and suppose that $\phi_\alpha : U_\alpha \to V_\alpha$ is a chart on M about q. Without loss of generality, suppose that $\phi_\alpha(0) = q$. Suppose that M is modelled on R^n and let $(\,,\,)$ and $|\,.\,|$ denote the Euclidean inner product and norm respectively on R^n. Then there exists $\delta > 0$ so that

$$U \equiv \{x \in R^n : |x| \leq \delta\} \subseteq U_\alpha$$

and $q \in V \equiv \phi_\alpha(U) \subseteq V_\alpha$. The first step in showing that d satisfies 4.1.6 (iv) is to examine its behaviour on V.

Given $u = [\phi_\alpha, u^o]_p$ and $v = [\phi_\alpha, v^o]_p$, where $p \in V_\alpha$, define a new Riemannian metric on V_α by

(4.1.1) $p \mapsto \langle\ ,\ \rangle_p^o$, where $\langle u, v\rangle_p^o = (u^o, v^o)$.

(In other words, the natural Riemannian metric on U_α has been mapped onto V_α via ϕ_α.) Denote $(\langle u, u\rangle_p^o)^{\frac{1}{2}}$ by $\|u\|_p^o$.

Define a map $\mu : V_\alpha \times S^{n-1} \to \mathbf{R}$ by $\mu(p, x) = \|[\phi_\alpha, x]_p\|_p$, where $S^{n-1} = \{x \in \mathbf{R}^n : |x| = 1\}$. Then μ is continuous and strictly positive and so, since $V \times S^{n-1}$ is compact, there exist $m, M > 0$ such that

(4.1.2) $m \le \mu(p, x) \le M$ on $V \times S^{n-1}$.

Let $[\phi_\alpha, y]_p \in T_p(M)$; then $y/|y| \in S^{n-1}$. Thus (4.1.2) is applicable and shows that

$$m\|[\phi_\alpha, y]_p\|_p^o = m|y| \le |y|.\ \|[\phi_\alpha, y/|y|]_p\|_p \le M\|[\phi_\alpha, y]_p\|_p^o.$$

But $|y|.\ \|[\phi_\alpha, y/|y|]_p\|_p = \|[\phi_\alpha, y]_p\|_p$ and so

(4.1.3) $m\|v\|_p^o \le \|v\|_p \le M\|v\|_p^o$

for all $p \in V$ and $v \in T_p(M)$.

Let $r \in M$ with $r \neq q$; we will show that $d(q, r) \neq 0$.

Case 1: $r \notin V$

Let $\xi : [a, b] \to M$ be a broken analytic curve with endpoints q, r. There exists $c \in [a, b]$ such that (i) $\xi(c)$ lies on the boundary of V at the point s, say, and (ii) the range of ξ on $[a, c]$ lies within V. In this case,

$$\begin{aligned} L_{q,r}(\xi) \ge L_{q,s}(\xi) &= \int_a^c \|\xi_{*,t}(1)\|_{\xi(t)} dt \\ &\ge m\int_a^c \|\xi_{*,t}(1)\|_{\xi(t)}^o dt \\ &= m\int_a^c |(\phi_\alpha^{-1}\xi)'(t)|\, dt. \end{aligned}$$

Now $\phi_\alpha^{-1}\xi$ on $[a, c]$ is a broken analytic curve in U from 0 to some point on its boundary and so its length is not less than δ. Hence

$$L_{q,s}(\xi) \geq m\delta > 0,$$

whereby $d(q, r) \neq 0$ as required.

Case 2: $r \in V$

Arguing as above we see that for each broken analytic curve $\xi : [a, b] \to M$ joining q and r,

$$L_{q,r}(\xi) \geq m|\phi_\alpha^{-1}(r)| > 0,$$

whence $d(q, r) \neq 0$ once again.

Thus d is indeed a metric.

To show that the metric topology corresponding to d is equivalent to the original topology of M, all that need be done is to show that to each point of M there corresponds a neighbourhood in which the two topologies are equivalent. But this has already been done for, from (4.1.3),

$$m\, d^o(x, y) \leq d(x, y) \leq M\, d^o(x, y)$$

for all points $x, y \in V$, where d^o denotes the metric on V_α obtained from the Riemannian metric (4.1.1), and evidently the natural topology and the d^o-topology on V_α are equivalent. //

4.1.8 **Geodesics.** If $\xi : [a, b] \to M$ is an analytic curve from a bounded interval $[a, b]$ into a Riemannian manifold M, then ξ is said to be a geodesic if $L_{p,q}(\xi) = d(p, q)$, where p and q are the end-points of ξ.

In the strict sense two curves are different even if they have the same domain but arise from different functions. However if the curves (that is, functions) coincide after an analytic change of parameters, then they will often be taken as the same curve. This is especially appropriate for the case of geodesics since here our primary concern is length and we know from Exercise 4.B that this is invariant under analytic change of parameter. Once convenient parameter for a geodesic is length ... see Exercise 4.C.

Even apart from the case of a change of parameter, if two points are joined by a geodesic, then it is not true in general that this geodesic is unique. (Consider the geodesics joining $z = 1$ and $z = -1$ in T.) Perhaps a more fundamental consideration than that of uniqueness is the question of existence. It is not true in general that every pair of points in a connected Riemannian manifold can be joined by a geodesic (consider the points $(-1, 0)$ and $(1, 0)$ in $\mathbf{R}^2 \setminus \{0\}$) but it is true whenever the manifold is also complete. The remainder of this section will be taken up by a sketch of this result; for omitted details the reader is referred to Helgason [1, Chapter 1, §10] and Kobayashi and Nomizu [1, Chapter IV, §4].

Suppose that M is a complete connected Riemannian manifold with associated metric d. For each p in M and $r > 0$, let

$$B_r(p) = \{q \in M : d(p, q) < r\}$$

and

$$E_r(p) = \{q \in \overline{B_r(p)} : p \text{ and } q \text{ can be joined by a geodesic}\}.$$

Clearly $\overline{B_r(p)}$, the closure of $B_r(p)$ in M, is equal to $\{q \in M : d(p, q) \leq r\}$. It follows from Exercise 4.D that for each p in M there always exists $\varepsilon > 0$ such that $\overline{B_\varepsilon(p)} = E_\varepsilon(p)$. Moreover the references cited above show that both $\overline{B_r(p)}$ and $E_r(p)$ are compact.

4.1.9 **Theorem.** <u>Every pair of points in a complete connected Riemannian manifold can be joined by a geodesic.</u>

Proof. Let $p \in M$; the method of proof will be to show that

$$\text{(4.1.5)} \quad \overline{B_r(p)} = E_r(p)$$

for all $r > 0$. As stated above, there exists $r > 0$ such that (4.1.5) is satisfied - let R be the supremum of all such r's. If $R = \infty$ there is nothing to prove. If on the contrary, R is finite, then (4.1.5) is satisfied for $r = R$. (Let $q \in \overline{B_R(p)}$; then q is the limit of a sequence of points each of which belongs to some $\overline{B_r(p)} = E_r(p)$, where $r < R$, and hence to $E_R(p)$. Since $E_R(p)$ is closed, the limit q of the sequence also belongs to $E_R(p)$.) We will show that (4.1.5) is satisfied for $r = R + \rho$, where ρ is a strictly positive real number.

By compactness of $\overline{B_R(p)} = E_R(p)$ there exist finitely many points $p_1, \ldots, p_k$ in $E_R(p)$ and positive numbers $\rho_1, \ldots, \rho_k$ such that

(i) the open balls $B_{\rho_j}(p_j)$ $(1 \leq j \leq k)$ cover $E_R(p)$, and

(ii) each ball is relatively compact and satisfies $\overline{B_{\rho_j}(p_j)} = E_{\rho_j}(p_j)$. Since the union of these balls is open and relatively compact, it is metric bounded and its complement is closed. Thus there exists a point in the complement which is at a minimum distance from p. This distance is strictly greater than R and so there must exist a number ρ satisfying $0 < \rho < \min\{\rho_1, \ldots, \rho_k\}$ and $\overline{B_{R+\rho}(p)} \subseteq \cup_{j=1}^{k} B_{\rho_j}(p_j)$.

Suppose now that q is a point in M such that $R < d(p, q) \leq R + \rho$. Let z be any point on the sphere $\{x : d(p, x) = R\}$ which is at minimum distance from q. Since every analytic curve joining p and q must intersect this sphere, it follows that

$$d(p, q) = d(p, z) + d(z, q).$$

Thus $d(z, q) = d(p, q) - d(p, z) < R + \rho - R = \rho$ and so from the manner in which ρ was selected, z and q lie in the same ball in the above cover of $E_R(p)$, $B_{\rho_i}(p_i)$ say. Combining the geodesic joining p and z with the geodesic joining z and q (this second geodesic exists since $B_{\rho_i}(p_i)$ satisfies condition (ii)) we obtain a curve of length $d(p, q)$ joining p and q. Thus $\overline{B_{R+\rho}(p)} = E_{R+\rho}$, contradicting the finiteness of R. //

4.2 Invariant metrics on Lie groups

It is a simple matter to define a Riemannian metric on a Lie group. Let G be a Lie group modelled on E and equip $T_e(G) = \underline{\underline{g}} = E$ with an inner product $\langle\ ,\ \rangle_e$. For each x in G define an inner product on $T_x(G)$ by

(4.2.1) $\langle X, Y\rangle_x = \langle (L_{x^{-1}})_{*,x}X, (L_{x^{-1}})_{*,x}Y\rangle_e$,

where $X, Y \in T_x(G)$. Some tedious manipulations similar to those used in the proof of Lemma 2.2.1 show that the family of inner products $x \mapsto \langle\ ,\ \rangle_x$, $x \in G$, is analytic in the sense of Definition 4.1.1 and hence is a Riemannian metric.

If a metric satisfies (4.2.1) then it also satisfies

(4.2.2) $\langle (L_y)_{*,x}X, (L_y)_{*,x}Y\rangle_{L_y x} = \langle X, Y\rangle_x$

for all $x, y \in G$ and $X, Y \in T_x(G)$. Condition (4.2.2) implies precisely that L_y is an isometry of G for all y in G.

4.2.1 **Definition.** Let M and N be two Riemannian manifolds. If $f : M \to N$ is a bijection such that both f and f^{-1} are analytic and if f also satisfies

$$\langle f_{*,p}X, f_{*,p}Y\rangle_{f(p)} = \langle X, Y\rangle_p$$

for all p in M and $X, Y \in T_p(M)$, then f is said to be an <u>isometry.</u>

It is a simple matter to verify that if M and N are both connected Riemannian manifolds, and that if their corresponding metrics are d_M and d_N, then each isometry $f : M \to N$ is distance preserving, that is,

$$d_M(p, q) = d_N(f(p), f(q))$$

for all p, q in M. (For ξ is a broken analytic curve joining p and q if and only if $f \circ \xi$ is a broken analytic curve joining $f(p)$ and $f(q)$, and every broken analytic curve joining $f(p)$ and $f(q)$ is of this form.) On the other hand, Theorem 11.1 of Chapter 1 of Helgason [1] shows that every distance preserving map of a connected Riemannian manifold onto itself is an isometry, but we will have no need of this fact here.

Returning to the case of a Lie group G, if a Riemannian metric on G is such that $L_x : G \to G$ is an isometry for all x in G (that is, if formula (4.2.2) is satisfied), then this metric is said to be <u>left invariant.</u> Similarly, if R_x is an isometry for all x in G, then the metric is said to be <u>right invariant.</u> A Riemannian metric which is both left and right invariant is said to be invariant.

The remainder of this section will be taken up with establishing a useful necessary and sufficient condition for a Lie group to possess an invariant Riemannian metric. As a corollary to this result it will be seen that all compact groups possess invariant Riemannian metrics.

4.2.2 **Adjoint representation.** Let G be a Lie group and for each x in G define the analytic automorphism

$$A_x : g \mapsto xgx^{-1}$$

on G. Next define the operator Ad from G into the set of linear maps from $\underline{\underline{g}}$ into $\underline{\underline{g}}$ by

$$\mathrm{Ad} : x \mapsto (A_x)_{*,e} \quad \text{for} \quad x \in G.$$

Since $(A_{xy})_{*,e} = (A_xA_y)_{*,e} = (A_x)_{*,e}(A_y)_{*,e}$ and $(A_{x^{-1}})_{*,e} = (A_x)_{*,e}^{-1}$, Ad is a representation of G by linear operators in $GL(\underline{\underline{g}})$ - it is called the adjoint representation of G. Since A_x can be written as $L_xR_{x^{-1}}$, for each X in $\underline{\underline{g}}$ we can write

$$\mathrm{Ad}(x)(X) = (L_x)_{*,x^{-1}}(R_{x^{-1}})_{*,e}(X)$$

and it follows from the techniques used in the proof of Lemma 2.2.1 that $x \mapsto \mathrm{Ad}(x)(X)$ is analytic on G for each X in $\underline{\underline{g}}$. In particular the adjoint representation is continuous in the sense of the definition in section A.2 of the Appendix.

4.2.3 **Theorem.** A Lie group G possesses an invariant Riemannian metric if and only if $\mathrm{Ad}(G)$ is relatively compact in $GL(\underline{\underline{g}})$.

Proof. Suppose that G possesses an invariant Riemannian metric $x \mapsto \langle\ ,\ \rangle_x$. Since it is right invariant we must have

$$(4.2.3)\quad \langle (R_{x^{-1}})_{*,e}X,\ (R_{x^{-1}})_{*,e}Y\rangle_{x^{-1}} = \langle X,\ Y\rangle_e$$

for all $x \in G$ and $X, Y \in T_e(G) = \underline{\underline{g}}$. Furthermore, since it is also left invariant

$$(4.2.4)\quad \langle (L_x)_{*,x^{-1}}X_0, (L_x)_{*,x^{-1}}Y_0\rangle_e = \langle X_0, Y_0\rangle_{x^{-1}}$$

for all $x \in G$ and $X_0, Y_0 \in T_{x^{-1}}(G)$. Combining (4.2.3) and (4.2.4) by putting $X_0 = (R_{x^{-1}})_{*,e}X$ and $Y_0 = (R_{x^{-1}})_{*,e}Y$ and recalling that $Ad(x) = (L_x)_{*,x^{-1}}(R_{x^{-1}})_{*,e}$ yields

$$\langle Ad(x)X, Ad(x)Y\rangle_e = \langle X, Y\rangle_e.$$

Since $\langle\ ,\ \rangle_e$ is an inner product on $\underline{g}$ we have that $Ad(x)$ is an orthogonal operator in $GL(\underline{g})$ for each x in G. Thus $Ad(G)$ is a subset of the orthogonal group $0(\underline{g}, \langle\ ,\ \rangle_e)$ and since this group is compact (Example 2.1.2 (c)), the same is true of the closure of $Ad(G)$.

Suppose now that $Ad(G)$ is relatively compact. Then the closure of $Ad(G)$, H say, is a compact subgroup of $GL(\underline{g})$ and so possesses an invariant normalised Haar measure, λ say. (See section A.4 of the Appendix.) Let $(\ ,\ )$ denote an inner product on $T_e(G)$. (This is possible since $T_e(G)$ is finite-dimensional.) Define a new linear product by

$$\langle X, Y\rangle_e = \int_H (s(X), s(Y))d\lambda(s)$$

for X, Y in $T_e(G)$. (Check that this does indeed define an inner product.) Given $t \in H$, then from the invariance of the Haar measure λ:

$$\begin{aligned}\langle t(X), t(Y)\rangle_e &= \int_H (st(X), st(Y))d\lambda(s)\\ &= \int_H (s(X), s(Y))d\lambda(s) = \langle X, Y\rangle_e.\end{aligned}$$

In particular, $\langle Ad(x)X, Ad(x)Y\rangle_e = \langle X, Y\rangle_e$ for all $x \in G$ and $X, Y \in T_e(G)$.

As was done in (4.2.1), define a left invariant Riemannian metric on G based on $\langle\ ,\ \rangle_e$. To show that this metric is also right invariant it suffices to show that

$$\langle (R_{x^{-1}})_{*,e}X, (R_{x^{-1}})_{*,x}Y\rangle_e = \langle X, Y\rangle_x$$

for all $x \in G$ and $X, Y \in T_x(G)$. Now this is immediate since

$$\langle (R_{x^{-1}})_{*,x}X, (R_{x^{-1}})_{*,y}Y\rangle_e = \langle \mathrm{Ad}(x^{-1})(R_{x^{-1}})_*X, \mathrm{Ad}(x^{-1})(R_{x^{-1}})_*Y\rangle_e$$
$$= \langle (L_{x^{-1}})_*X, (L_{x^{-1}})_*Y\rangle_e$$
$$= \langle X, Y\rangle_x$$

using first the invariance of $\langle\ ,\ \rangle_e$ under the adjoint representation and then the invariance of the metric under left translations. //

4.2.4 **Corollary.** A Riemannian metric $x \mapsto \langle\ ,\ \rangle_x$ on a Lie group G is invariant if and only if

$$\langle \mathrm{Ad}(x)X, \mathrm{Ad}(x)Y\rangle_e = \langle X, Y\rangle_e$$

for all $x \in G$ and $X, Y \in T_e(G)$.

The adjoint representation is (uniformly) continuous from G into $GL(\underline{\underline{g}})$ and so if G is compact, then the same is true of $\mathrm{Ad}(G)$. Thus:

4.2.5 **Corollary.** Every compact group possesses an invariant Riemannian metric.

4.3 Geodesics on Lie groups

If M is an analytic manifold and $\xi : [a, b] \to M$ is a curve on M, then $\xi : [c, d] \to M$ is said to be a segment of this curve if $a \le c \le d \le b$. In the case of R^n it is obvious that a curve is a geodesic if and only if it is a translate of a segment of a 1-parameter subgroup. In this section we will show that, subject to certain restrictions, this result is valid for all Lie groups which possess invariant Riemannian metrics. This will enable us to show that for all such groups which are also complete and connected, in particular for connected compact Lie groups, the exponential maps are surjective.

In general it is not true that a segment of a 1-parameter subgroup is a geodesic. For example, the curve $\theta : t \mapsto e^{ikt}$, $k \in Z$, from $[0, 2\pi]$ into T is a segment of a 1-parameter subgroup of T but it is certainly not a geodesic when $k \neq 0$. However it is a local geodesic:

4.3.1 **Local geodesics.** A curve $\xi : [a, b] \to M$ from a bounded interval $[a, b]$ into a Riemannian manifold M is said to be a local geodesic if the range of the curve can be covered by open sets such that in each of these sets every segment of the curve is a geodesic.

In the other direction, for a geodesic to be a translation of a segment of a 1-parameter subgroup we would expect that at the very least a change of parameter will be needed. For example, $\xi : [0, 1] \to \mathbf{R}$ defined by $\xi(t) = t^2$ is a geodesic joining 0 and 1 but it is not a translation of a segment of any 1-parameter subgroup of **R**. However ξ is coincident with a segment of the 1-parameter subgroup $\theta : t \mapsto t$:-

4.3.2 **Coincident curves.** Two broken analytic curves in an analytic manifold are said to be coincident if their ranges are equal.

It follows from Exercise 4.B that the lengths of coincident injective curves are equal and hence that an injective curve is a geodesic if and only if each injective curve coincident with it is also a geodesic.

4.3.3 **Theorem.** Let G be a Lie group with an invariant Riemannian metric. Then every geodesic in M is coincident with a translation of a segment of a 1-parameter subgroup and (almost conversely) every such translation is a local geodesic.

The following lemma is needed to prove the first part of this theorem.

4.3.4 **Lemma.** Let σ be an isometry of a Riemannian manifold M with σ^2 the identity operator. Let x be an isolated fixed point of σ and let U be a neighbourhood of x such that (i) every pair of points in U can be joined by a unique (up to coincident curves) geodesic (see Exercise 4.D) and (ii) U contains no other fixed points of σ. If y and $\sigma(y)$ belong to U, the geodesic joining y and $\sigma(y)$ passes through x.

Proof. Let ξ be the unique geodesic joining y and $\sigma(y)$. Since $\sigma^2(z) = z$ and σ is an isometry, $\sigma \circ \xi$ is the geodesic joining $\sigma(y)$ and y. Thus σ is a continuous map of the range of ξ into itself and hence, from a basic fixed-point theorem, there exists a point on the range of ξ which is a fixed point for σ; this point can only be x. //

Proof (of Theorem 4. 3. 3). Let G be a Lie group with an invariant Riemannian metric $x \mapsto \langle\, ,\, \rangle_x$. Let U be a convex symmetric open neighbourhood of 0 in $\underline{g}$ on which the exponential map is an analytic homeomorphism and let W be a symmetric neighbourhood of e in G such that (i) $W^2 \subseteq \exp \frac{1}{2}U$ and (ii) every pair of points in W can be joined by a unique geodesic lying within W. Choose $\alpha > 0$ in such a way that $\exp(\alpha U) \subseteq W$. We introduce one more set: since the topology of G can be defined by the metric d derived from the Riemannian metric (Theorem 4. 1. 7), then $\exp(\alpha U)$ must contain a set of the form $B_\rho = \{x : d(e, x) \leq \rho\}$, $\rho > 0$. In all,

(4. 3. 1) $$e \in B_\rho \subseteq \exp(\alpha U) \subseteq W; \quad W^2 \subseteq \exp \tfrac{1}{2}U.$$

Let $\xi : [a, b] \to G$ be a geodesic; without loss of generality we suppose that $\xi(a) = e$. If $\operatorname{ran}(\xi) \cap \{x : d(e, x) = \rho\}$ is nonempty, let a_1 denote the greatest number not exceeding b such that the curve ξ on $[a, a_1]$ lies entirely within B_ρ and thus $z_1 = \xi(a_1)$ satisfies $d(e, z_1) = \rho$; otherwise set $z_1 = \xi(b)$. Define $\theta_1 : [0, 1] \to G$ by $\theta_1(t) = \exp tZ_1$ where $Z_1 = \log z_1$; we will show that θ_1 and $\xi_1 = \xi|_{[a, a_1]}$ are coincident.

First of all, from (4. 3. 1) notice that $\operatorname{ran}(\theta_1) \subseteq W$ and in particular, $x = \theta_1(\frac{1}{2}) \in W$. Define $\sigma : G \to G$ by $\sigma(y) = xy^{-1}x$. To show that σ is an isometry of G, we first show that this is true of $\lambda : y \to y^{-1}$. For $p \in G$ and $X, Y \in T_p(G)$,

$$\begin{aligned}\langle \lambda_{*,p}X, \lambda_{*,p}Y\rangle_{\lambda(p)} &= \langle (L_p)_{*,p^{-1}}\lambda_{*,p}X, (L_p)_{*,p^{-1}}\lambda_{*,p}Y\rangle_e \\ &= \langle (L_p \circ \lambda)_* X, (L_p \circ \lambda)_* Y\rangle_e \\ &= \langle (\lambda \circ R_{p^{-1}})_* X, (\lambda \circ R_{p^{-1}})_* Y\rangle_e \\ &= \langle \lambda_{*,e}(R_{p^{-1}})_{*,p}X, \lambda_{*,e}(R_{p^{-1}})_{*,p}Y\rangle_e,\end{aligned}$$

where the first line follows from the left invariance of the Riemannian metric. Working on a neighbourhood of 0 in $\underline{g}$ on which $\exp$ is an analytic homeomorphism, we have $\log \circ \lambda \circ \exp(X) = -X$ and hence

$$\lambda_{*,e}[\exp, v]_e = [\exp, (\log \circ \lambda \circ \exp)'(0)(v)]_e$$
$$= [\exp, -v]_e.$$

Thus

$$\langle \lambda_{*,e}(R_{p^{-1}})_{*,p}X, \lambda_{*,e}(R_{p^{-1}})_{*,p}Y \rangle_e$$
$$= (-1)^2 \langle (R_{p^{-1}})_{*,p}X, (R_{p^{-1}})_{*,p}Y \rangle_e$$

which is equal to $\langle X, Y \rangle_p$ by the right invariance of the Riemannian metric. Combining these manipulations results in

$$\langle \lambda_{*,p}X, \lambda_{*,p}Y \rangle_{\lambda(p)} = \langle X, Y \rangle_p,$$

whence the fact that λ is an isometry. Since $\sigma = L_x \circ R_x \circ \lambda$, we have that σ is an isometry as required.

Now $\sigma^2(y) = x(xy^{-1}x)^{-1}x = y$ for all y in G and also x is a fixed point of σ. The next step is to show that x is the only fixed point of σ in W. If $y \in W$ and $\sigma(y) = y$, then $(xy^{-1})^2 = e$. Reading from (4.3.1) we know that there exists $X \in \frac{1}{2}U$ such that $\exp X = xy^{-1}$ and hence $\exp(2X) = (xy^{-1})^2 = e$. But $2X \in U$, a set on which $\exp$ is a bijection, and so $X = 0$. Thus $xy^{-1} = e$, that is $x = y$.

The previous lemma is now applicable and shows that the geodesic $\xi|_{[a, a_1]}$ joining $\sigma(z_1) = \theta_1(\frac{1}{2})\theta_1(-1)\theta_1(\frac{1}{2}) = e$ and z_1 passes through the point x. By repeating this argument for the two half segments of the 1-parameter subgroup $t \mapsto \exp tZ_1$ on $[0, \frac{1}{2}]$ and $[\frac{1}{2}, 1]$, and then for the quarter segments, and so on, it follows that $\theta_1(m.\,2^{-n}) \in \mathrm{ran}(\xi_1)$ for every $n \in Z^+$ and $0 \le m \le 2^n$. By continuity $\theta_1(t)$ belongs to $\mathrm{ran}(\xi_1)$ for all $t \in [0, 1]$ and so, since ξ_1 is a geodesic, ξ_1 and θ_1 are coincident.

If $z_1 = \xi(b)$, then the proof is complete. Suppose $z_1 \ne \xi(b)$ and consider the geodesic $\xi^1 : [a_1, b] \to G$ defined by $\xi^1(t) = \xi(t)$ for $t \in [a_1, b]$. First note that the length of ξ^1 is shorter than the length of ξ by at least an amount ρ. By examining the geodesic $L_{\xi(a_1)^{-1}} \circ \xi^1$ we can repeat the above argument to find $Z_2 \in \underline{g}$ and $a_2 \in [a_1, b]$ such that $\xi_2 = L_{\xi(a_1)^{-1}} \circ \xi^1[a_1, a_2]$ and $\theta_2 : [0, 1] \to G$, where $\theta_2(t) = \exp(tZ_2)$, are coincident. Since each time we repeat this step we shorten ξ by at least ρ, this process must terminate after a finite number of steps. Thus there exists an integer m, a decomposition $a < a_1 < \ldots < a_{m-1} < a_m = b$ and a set $\{Z_1, \ldots, Z_m\} \subseteq \underline{g} \setminus \{0\}$ such that $\xi : [a, b] \to G$ and $\theta : [0, m] \to G$ are coincident, where

$$\theta(t) = \begin{cases} \exp(tZ_1) & \text{for } 0 \leq t \leq 1 \\ \xi(a_1).\exp((t-1)Z_2) & \text{for } 1 \leq t \leq 2 \\ \cdots\cdots\cdots\cdots & \\ \cdots\cdots\cdots\cdots & \\ \xi(a_{m-1})\exp((t-m+1)Z_m) & \text{for } m-1 \leq t \leq m. \end{cases}$$

Since θ is at least continuous, evaluation of θ at the points $1, 2, \ldots, m-1$ yields

$$(4.3.2) \quad \xi(a_k) = \exp(Z_1) \ldots \exp(Z_k) \quad \text{for } k \in \{1, 2, \ldots, m\}.$$

Suppose we have shown also that for each k in $\{1, 2, \ldots, m-1\}$ there exists a nonzero real number λ_k such that

$$(4.3.3) \quad Z_{k+1} = \lambda_k Z_k.$$

Then, combining this supposition with (4.3.2) shows, for $k \leq t \leq k+1$, that

$$\theta(t) = \xi(a_k)\exp((t-k)Z_{k+1})$$
$$= \exp(\mu_{k-1}Z_1).\exp((t-k)(\mu_k - \mu_{k-1})Z_1),$$

(where $\mu_r = 1 + \lambda_1 + \lambda_1\lambda_2 + \ldots + \lambda_1 \ldots \lambda_r$ for $r \geq 1$ and 0 otherwise)

$$= \exp(\{\mu_{k-1} + (t-k)(\mu_k - \mu_{k-1})\}Z_1).$$

Make the change of parameter $u = \mu_{k-1} + (t-k)(\mu_k - \mu_{k-1})$ and we have that θ on $[k, k+1]$ is coincident with $t \mapsto \exp(tZ_1)$ on $[\mu_{k-1}, \mu_k]$ and hence that θ on $[0, m]$ is coincident with $t \mapsto \exp(tZ_1)$ on $[0, \mu_{m-1}]$. Thus ξ on $[a, b]$ is coincident with a segment of a 1-parameter subgroup as required.

It remains to show the validity of (4. 3. 3). First note that since $\frac{d}{dt}\{x. \exp tX\}\big|_{t=k} = X(x. \exp kX)$, then

$$\frac{d}{dt}\{\xi(a_{k-1})\exp((t - k + 1)Z_k)\}\big|_{t=k} = Z_k(\xi(a_{k-1})\exp Z_k)$$

and

$$\frac{d}{dt}\{\xi(a_k)\exp((t - k)Z_{k+1})\}\big|_{t=k} = Z_{k+1}(\xi(a_k)).$$

(Here we have used Z_k and Z_{k+1} to denote both the elements in $T_e(G)$ and their corresponding left invariant vector fields.) Since θ is a geodesic and since both the above derivatives are different from zero (if a left invariant vector field vanishes at a point, it vanishes everywhere!), Exercise 4. E shows that $\theta_{*,t}(1)$ can only change by a nonzero linear multiple at each of the points 1, 2, ..., m - 1. In other words, for each $k \in \{1, 2, \ldots, m - 1\}$ there exists nonzero $\lambda_k \in \mathbf{R}$ such that

$$Z_{k+1}(\xi(a_k)) = \lambda_k . Z_k(\xi(a_{k-1}). \exp Z_k).$$

However (4. 3. 2) shows that $\xi(a_k) = \xi(a_{k-1}). \exp Z_k$ and since a left invariant vector field is fully determined by its value at any given point, the validity of (4. 3. 3) follows.

For the second part of the proof consider the 1-parameter subgroup $t \mapsto \exp tX$ on $[a, b]$ with $X \neq 0$. Since the metric is invariant, in considering whether or not this segment is a local geodesic it suffices to consider the same question for $t \mapsto L_{\exp(-aX)}\exp tX = \exp((t - a)X)$ $t \in [a, b]$, that is, for $\theta : t \mapsto \exp tX$ on $[0, b - a]$. In the usual proof of Exercise 4. D it is shown that corresponding to every $X \in T_e(G)$ there exist $c \in G$ (with $c \neq e$) and a geodesic $\xi : [0, 1] \to G$ such that $\xi(0) = e$, $\xi(1) = c$ and $\xi_{*,0}(1) = X$. (See also the Notes at the end of this Chapter.) From the first part of this result, ξ is coincident with a segment of a nontrivial 1-parameter subgroup, $\psi : t \to \exp tY$ with $t \in [0, d]$, say.

Now ξ and ψ are coincident in a neighbourhood of the identity and so, in particular, their derivatives at the identity are equal up to a scalar factor. Thus X is equal to a nonzero scalar multiple of Y and so θ and ψ are coincident in a neighbourhood of e. But this completes the proof since ψ and ξ are also coincident in a neighbourhood of e and furthermore ξ is a geodesic. //

4.3.5 **Corollary.** Let G be a complete connected Lie group which possesses an invariant Riemannian metric. (In particular, let G be a connected compact Lie group.) Then the associated exponential map is surjective.

Proof. Suppose that G satisfies the hypotheses of the Corollary and that x is an element of G different from its identity. From Theorem 4.1.9 there exists a geodesic joining e and x, and from the proceding theorem this geodesic is of the form $t \mapsto \exp tX$, where $t \in [0, b]$ and $X \in \underline{g}$. In particular, $x = \exp bX$ and so exp is surjective, as asserted. //

4.3.6 **Application.** Let G be a Lie group with a left invariant Riemannian metric $x \mapsto \langle \, , \rangle_x$. It is possible to explicitly determine the lengths of segments of 1-parameter subgroups of G in terms of this Riemannian metric. In fact, the length of $\theta : t \mapsto \exp tX$ for $X \in \underline{g}$ and $t \in [a, b]$ is given by $(b - a)\|X\|_e$. (The full formula for the length of θ is $\int_a^b \|\theta_{*,t}(1)\|_{\theta(t)} dt$ which is equal to $\int_a^b \|(L_{\theta(t)^{-1}})_{*,\theta(t)}\theta_{*,t}(1)\|_e dt$ since the metric is left invariant. Now

$$\begin{aligned}(L_{\theta(t)^{-1}})_{*,\theta(t)}\theta_{*,t}(1) &= (L_{\theta(t)^{-1}})_{*,\theta(t)}X(\theta(t)) \\ &= X(L_{\theta(t)^{-1}}\theta(t)) = X(e),\end{aligned}$$

the first line following from the left invariance of X.)

Notes

In the case of R^n it is clear that a curve is a geodesic if and only if it is coincident with a curve which has the properties that its derivative never vanishes and that all its tangent vectors are parallel. In a general differentiable manifold it is not usually possible to immediately compare tangent vectors at different points. However it is possible in the case of a manifold possessing a linear connection, and hence for all Riemannian manifolds (see Helgason [1, 1. §4 and 1. §9]). A curve in a Riemannian manifold M which has 'parallel tangents' will be referred to as a parallel geodesic and it is known that around each point in M there exists a neighbourhood in which a curve is a geodesic if and only if it is a parallel geodesic.

Furthermore, the fundamental existence and uniqueness theorem for parallel geodesics in a Riemannian manifold M is an extension of the obvious result in $\mathbf{R}^n$. Namely, corresponding to each point $p \in M$ there is a neighbourhood U of p and a real number $\varepsilon > 0$ such that for every $q \in U$ and every tangent vector $X \in T_q(M)$ with $\|X\|_q < \varepsilon$, there is a unique parallel geodesic $\gamma_X : [-1, 1] \to M$ satisfying

$$\gamma_X(0) = q, \quad (\gamma_X)_{*,0}(1) = X.$$

(For a proof see Helgason [1, p. 30, Proposition 5.3] or Spivak [1, p. 9-45].) The method of proof is a generalisation of the proof of Theorem 2.2.3 (ii) since it involves associating with each geodesic a certain differential equation and then showing that this differential equation has a unique solution.

From here it is possible to introduce a notion of an 'exponential' map for a Riemannian manifold M which extends that introduced in Chapter 2 for Lie groups. If $X \in T_q(M)$ is a vector for which there is a parallel geodesic $\gamma_X : [-1, 1] \to M$ satisfying

$$\gamma_X(0) = q, \quad (\gamma_X)_{*,0}(1) = X$$

(and this is always the case in a suitable neighbourhood of 0 in $T_q(M)$), then we define the <u>exponential at</u> q <u>of</u> X to be

$$\exp_q(X) = \gamma_X(1).$$

The geodesic γ_X can then be described as

$$t \mapsto \gamma_X(t) = \exp_q(tX), \quad t \in [-1, 1].$$

Part of the reason for this terminology is that if the tangent space to $T_q(M)$ at 0 is identified with $T_q(M)$, then the map

$$(\exp_q)_{*,0} : T_q(M) \to T_q(M)$$

is the identity. (Compare with the proof of Theorem 2.2.12 where it is shown that $\exp_{*,0}$ is the identity map from $\underline{\underline{g}}$ into $\underline{\underline{g}}$ where G is a Lie group.) Since $(\exp_q)_{*,0}$ is linear, to see this fact we need only check that it is the identity map in some neighbourhood of zero; choose this neighbourhood to satisfy the conclusion of the existence and uniqueness theorem for geodesics as described above. Suppose that $X \in T_q(M) = T_0(T_q(M))$ is in this neighbourhood. Then $c : t \mapsto tX$ is a parallel geodesic in $T_q(M)$ with $c_{*,0}(1) = X$ and $c(0) = 0$. Hence

$$\gamma_X(t) = \exp_q(tX) = \exp_q \circ c(t)$$

and so

$$\begin{aligned}(\exp_q)_{*,0}(X) &= (\exp_q)_{*,0} \circ c_{*,0}(1)\\ &= (\exp_q \circ c)_{*,0}(1)\\ &= (\gamma_X)_{*,0}(1) = X.\end{aligned}$$

Furthermore, in the case of a Lie group $\exp_e$ is equal to the usual exponential map. For if G is a Lie group and $\exp_e$ is defined at $X \in T_e(G)$, then $\gamma_X : t \mapsto \exp_e(tX)$, $t \in [-1, 1]$, is a parallel geodesic with $(\gamma_X)_{*,0}(1) = X$ and $\gamma_X(0) = e$. However, $\theta : t \mapsto \exp(tX)$ is also a parallel geodesic with $\theta_{*,0}(1) = X$ and $\theta(0) = e$ and so, by the uniqueness result quoted above, $\exp(X) = \exp_e(X)$.

In the special case where G is a connected Abelian Lie group we saw from Exercise 3.B that G is of the form $T^m \times R^n$. It is clear that the metric associated with each invariant Riemannian metric on G is a scalar multiple of the usual metric for $T^m \times R^n$. Furthermore, all the results of sections 4.2 and 4.3 are trivial in this case. For example, note that the conclusion of Corollary 4.3.5, an important result, was only the subject of a minor exercise, Exercise 3.B (iii), when G was connected and Abelian.

The Lie groups covered by the hypotheses of Corollary 4.3.5 do not appear to be the only ones which have surjective exponential maps. For example, Exercise 4.F below shows that this property also belongs to the exponential maps of connected Lie groups with nilpotent Lie algebras.

Exercises

4. A. Use the description of analytic vector fields on $\mathbf{R}^n$ obtained in Exercise 2. A (i) to show that the construction given in 4. 1. 4 does indeed yield a Riemannian metric on $\mathbf{R}^n$.

4. B **Invariance of length.** Suppose that M is a Riemannian manifold and that $\xi : [a, b] \to M$ and $\eta : [c, d] \to M$, where $[a, b]$ and $[c, d]$ are bounded, are two analytic injective curves in M. Suppose further that they are coincident.

(i) If there exists a diffeomorphism $\theta : [a, b] \to [c, d]$ such that $\xi = \eta \circ \theta$, prove that

$$(\ast) \quad L_{\xi(a),\, \xi(b)}(\xi) = L_{\eta(c),\, \eta(d)}(\eta).$$

(ii) Prove that $(\ast)$ is valid in the general case. (Hint: Show that $\xi_{*,t}(1)$ and $\eta_{*,t}(1)$ are zero at a finite number of places at most. Now break the range of the curve into a finite number of segments on which both $\xi_{*,t}(1)$ and $\eta_{*,t}(1)$ are nonzero except perhaps at the end-points. The inverse mapping theorem and part (i) above can now be used to show that, apart from arbitrarily small neighbourhoods of these exceptional points, the lengths of these segments are the same whether considered as segments of ξ or η.)

4. C **Reparametrization by length.** Let ξ be an injective analytic curve from a bounded interval $[a, b]$ into a Riemannian manifold M. Suppose also that $\xi_{*,t}(1) \neq 0$ on $[a, b]$. For $t \in [a, b]$, define $s(t) = L_{\xi(a),\, \xi(t)}(\xi)$ and show that s is a diffeomorphism from $[a, b]$ onto $[0, L]$, where L is the length of ξ. Now define $\eta : [0, L] \to M$ by $\eta = \xi \circ s^{-1}$ and prove that ξ and η are coincident with

$$L_{\eta(0),\, \eta(t)}(\eta) = t \quad \text{for } t \in [0, L].$$

4. D. Show that around every point in a Riemannian manifold there exists a neighbourhood such that each pair of points in this neighbourhood can be joined by exactly one geodesic within this neighbourhood. (See Helgason [1, Chapter 1, Theorem 6. 2 and Lemma 9. 3]. This is a deep result and lies at the basis of the work of the preceding chapter. The following consideration will perhaps add plausibility to its statement. Since we are only concerned with local properties it is enough to consider the problem on an open subset of $\mathbf{R}^n$ which is equipped with a Riemannian metric. If the Riemannian metric is the usual metric on $\mathbf{R}^n$, then the problem is trivial. If not, then provided we work within a small enough subset the analyticity of the Riemannian metric will ensure that it is not 'far removed' from the usual $\mathbf{R}^n$ metric and that the geodesics sought after will be close to straight lines. Its formal proof is usually approached via the existence and uniqueness of parallel tangents geodesics described in the preceding Notes.)

4. E. Show that if $\xi : [0, 1] \to M$ and $\eta : [1, 2] \to M$ are analytic on a Riemannian manifold M and that if $\xi(1) = \eta(1)$ and the curve equal to ξ on $[0, 1]$ and η on $[1, 2]$ is a geodesic, then $\xi_{*,1}(1)$ is a real multiple of $\eta_{*,1}(1)$ provided $\eta_{*,1}(1) \neq 0$.

4. F. A Lie algebra L is said to be nilpotent if there exists a positive integer n such that

$$[X_1, [X_2, [\ldots [X_{n-1}, X_n] \ldots] = 0$$

for all $X_1, \ldots, X_n \in L$. Show that the exponential map is surjective whenever G is connected and $\underline{\underline{g}}$ is nilpotent. (Hint: There exists a neighbourhood U of 0 in $\underline{\underline{g}}$ and an analytic map $Z : U \times U \to \underline{\underline{g}}$ such that

$$\exp X \exp Y = \exp Z(X, Y)$$

for $X, Y \in U$. If $\underline{\underline{g}}$ is nilpotent, then the CBH formula shows that Z has an analytic extension to all of $\underline{\underline{g}} \times \underline{\underline{g}}$ and from Exercise 1. H we know that this extension is unique.)

5·Lie subgroups and subalgebras

The main purpose of this chapter is to introduce the notion of a Lie subgroup of a Lie group G and to show that a Lie group H is a Lie subgroup of G if and only if $\mathfrak{h}$ is a Lie subalgebra of $\mathfrak{g}$. This result will then be refined to provide a one-to-one correspondence between connected normal Lie subgroups of G and ideals of $\mathfrak{g}$. As part of the machinery needed to prove this second correspondence a fairly detailed analysis of the adjoint representation of a Lie group (first introduced in §4.2) will be given.

5.1 Subgroups and subalgebras

In this section we proceed to develop correspondences between certain subgroups of a Lie group and subalgebras of its Lie algebra.

5.1.1 **Lie subgroups.** If G and H are Lie groups and H is algebraically a subgroup of G, then H is said to be a *Lie subgroup* of G if the natural inclusion map $j : H \to G$ is continuous.

Examples of Lie subgroups are close at hand. For example, every closed subgroup of a Lie group is a Lie group when it is equipped with the induced topology and hence these closed subgroups are Lie subgroups. It is common in the definition of a Lie subgroup to also require that the inclusion map j is an analytic immersion (see 1.2.6). However this extra condition is redundant since its validity can be deduced from the above definition. As a matter of notation, the expression '$j : H \to G$ is a Lie subgroup of G' will mean precisely that H is a Lie subgroup of G and that j is the natural inclusion.

5.1.2 **Proposition.** *If* $j : H \to G$ *is a Lie subgroup of* G, *then:*

(i) j *is an analytic immersion; and*

(ii) $j_{*,e} : \mathfrak{h} \to \mathfrak{g}$ *is a Lie algebra monomorphism.*

Proof. Before commencing the proof we remark that the only properties of j used are that it is a continuous monomorphism from H into G. Since j is continuous, it follows from Theorem 2.3.1 that it is analytic. Now the definition of an immersion is that $j_{*,x}$ is an injection for each x in H. Exercise 5.A shows this to be a fact provided $j_{*,e}$ is an injection, which is the case if and only if $\{X \in \underline{\underline{h}} : j_{*,e}X = 0\} = \{0\}$.

Suppose that $X \in \underline{\underline{h}}$ satisfies $j_{*,e}X = 0$. Let U be a neighbourhood of 0 in $\underline{\underline{h}}$ on which $\exp_H$ is one-to-one and let t be a non-zero real number such that $tX \in U$. Since $j_{*,e}$ is linear, $j_{*,e}(tX) = 0$ and so it follows from Exercise 2.F that

$$j \circ \exp_H(tX) = \exp_G \circ j_{*,e}(tX) = e_G.$$

Thus $\exp_H(tX) = e_H$ and hence $tX = 0$. Therefore $X = 0$ and j is an analytic immersion as required. Furthermore $j_{*,e} : \underline{\underline{h}} \to \underline{\underline{g}}$ is a Lie algebra monomorphism since Exercise 2.G demonstrates that it is already a Lie homomorphism. //

Suppose that $j : H \to G$ is a Lie subgroup of G. In the sequel, when no confusion is likely, we will identify $\underline{\underline{h}}$ with its image in $\underline{\underline{g}}$ under $j_{*,e}$. The following result is the converse to Proposition 5.1.2 and its validity allows us to state that if G is a Lie group, then $H \mapsto \underline{\underline{h}}$ is a one-to-one correspondence between all the connected Lie subgroups H of G and all the Lie subalgebras $\underline{\underline{h}}$ of $\underline{\underline{g}}$.

5.1.3 **Theorem.** Let G be a Lie group with Lie algebra $\underline{\underline{g}}$. Corresponding to each Lie subalgebra L of $\underline{\underline{g}}$ there exists a unique connected Lie subgroup H of G such that $\underline{\underline{h}}$ is equal to L. (More precisely, corresponding to each Lie subalgebra L of $\underline{\underline{g}}$ there exists a unique connected Lie subgroup $j : H \to G$ of G such that $j_{*,e}(\underline{\underline{h}}) = L$.)

Proof. Let U be an open symmetric neighbourhood of 0 in $\underline{\underline{g}}$ such that (i) $\exp$ is an analytic homeomorphism on U (see Theorem 2.2.12) and (ii) $\log(\exp(X)\exp(Y)) \in L$ for all $X, Y \in U \cap L$ (see the CBH formula 3.2.6). Let H denote the subgroup of G generated by $\exp(U \cap L)$. The induced topology on H certainly makes H into a topological group, but it doesn't necessarily ensure that H has a compatible analytic atlas. To overcome this possible lack we proceed to give H a stronger topology under which it is always a Lie group. Define a basis of neighbourhoods about e in H by $\{\exp U' : U' \subseteq U$ and U' is an open neighbourhood of 0 in $L\}$. These neighbourhoods satisfy the requirements of Theorem A. 1. 2 and so the set of all left translates of these neighbourhoods forms an open subbase of a topology on H; under this topology H is a topological group. (To see that this topology is indeed stronger than the one induced from G, begin with a typical induced neighbourhood $H \cap W$ of e in H, where W is a neighbourhood of e in G. Now consider $U \cap L \cap \exp^{-1}W$; it is a neighbourhood of 0 in L so its image under exp must, by definition, be a neighbourhood of e in H. However

$$\exp(L \cap U \cap \exp^{-1}W) \subseteq (\exp(L \cap U)) \cap W \subseteq H \cap W.$$

Exercise 5. B contains an example where the subgroup topology is strictly stronger than the induced topology.)

Let $x = \exp X$, where $X \in L$. There exists $Y \in L \cap U$ and a positive integer k such that $kY = X$. Hence $\exp X = \exp kY = (\exp Y)^k$ and so $\exp X$ belongs to H. Thus H is also the subgroup of G generated by $\exp L$. Since L is connected and exp is continuous, then $\exp L$ is also connected. Let H_0 denote the connected component of e in H; then H_0 is a topological subgroup of H (Lemma 2. 1. 4 (i)). Since $\exp L$ is connected we must have $\exp L \subseteq H_0$. Thus the subgroup of G generated by $\exp L$, namely H, is a subset of H_0. Therefore $H = H_0$ showing that H is connected.

To see that H is a Lie group we seek to apply Theorem 2. 1. 3. The exponential map is a homeomorphism from $L \cap U$ onto some open subset V of H with $e \in V$ and so forms a chart about the origin. Furthermore, from the manner in which U was chosen we know that

$$(X, Y) \mapsto \log(\exp(X)\exp(Y))$$

is a map from $L \cap U$ into L. Certainly this map is analytic on $\{(X, Y) : X, Y \in L \cap U$ and $\exp X \exp Y \in \exp(L \cap U)\}$. Also

$$X \mapsto \log((\exp X)^{-1}) = -X$$

is analytic on $L \cap U$ so that, from the theorem mentioned above, H is a Lie group.

The following lemma will be used to complete the proof of 5.1.3.

5.1.4 **Lemma.** Let $j : H \to G$ be a Lie subgroup of G. Then $X \in j_{*,e}(\underline{\underline{h}})$ if and only if $t \mapsto j^{-1} \circ \exp_G tX$ is a well-defined continuous map from $\mathbf{R}$ into H.

Proof. Suppose that $X = j_{*,e}(Y)$, where $Y \in \underline{\underline{h}}$. Then

$$\exp_G tX = \exp_G j_{*,e}(tY)$$
$$= j \circ \exp_H tY \in j(H)$$

and so $t \mapsto j^{-1} \circ \exp_G tX = \exp_H tY$ is a well-defined analytic map from $\mathbf{R}$ into H. Conversely, suppose that $t \mapsto j^{-1} \circ \exp_G tX$ is a well-defined continuous map from $\mathbf{R}$ into H, where $X \in \underline{\underline{g}}$. Since this map is a homomorphism, it must be analytic, in fact a 1-parameter subgroup of H. From Theorem 2.2.10 there exists $Y \in \underline{\underline{h}}$ such that $j^{-1} \circ \exp_G tX = \exp_H tY$ for all t in $\mathbf{R}$. Thus

$$\exp_G tX = j \circ \exp_H tY = \exp_G \circ j_{*,e}(tY)$$

for all t in $\mathbf{R}$ and so $X = j_{*,e}(Y)$. //

Returning to the proof of Theorem 5.1.3, we will show that $j_{*,e}(\underline{\underline{h}}) = L$. Suppose that $X = j_{*,e}(Y)$, where $Y \in \underline{\underline{h}}$. From the above lemma, $j^{-1} \circ \exp(tX) \in H$ for all $t \in \mathbf{R}$ and so for t sufficiently small, $\exp(tX) \in \exp(L \cap U)$. Thus $X \in L$, whence $j_{*,e}(\underline{\underline{h}}) \subseteq L$. On the other hand, given $X \in L$, then $t \mapsto \exp_G tX$ is continuous from $\mathbf{R}$ into H. (For small t, $\exp_G tX \in \exp(U \cap L)$ and so $\exp_G tX \in H$ for all t. The manner in which we defined the topology of H implies that $t \mapsto j^{-1} \circ \exp_G tX$ is continuous.) From Lemma 5.1.4, $X \in j_{*,e}(\underline{\underline{h}})$ as required.

To prove uniqueness, suppose that $j_1 : H_1 \to G$ is another connected Lie subgroup of G which satisfies $(j_1)_{*,e}(\underline{\underline{h}}_1) = L$. Since $\exp_G(L) \subseteq j_1(H_1) \cap j(H)$ and both $j_1 H_1$ and jH are connected as subsets of G, all that is required is to show that $j_1^{-1} \circ j$ provides a homeomorphism on some neighbourhood of e in H. Let W be an open connected neighbourhood of 0 in L such that $\exp_{H_1}$ and $\exp_H$ are analytic homeomorphisms on $(j_1)_{*,e}^{-1}(W)$ and $j_{*,e}^{-1}(W)$ respectively. Evidently for each set $V \subseteq W$, $\exp_{H_1}(j_1)_{*,e}^{-1}(V)$ is open in H_1 if and only if $\exp_H \circ j_{*,e}^{-1}(V)$ is open in H, that is, $j_1^{-1} \circ \exp_G(V)$ is open in H_1 if and only if $j^{-1} \circ \exp_G(V)$ is open in H. This completes the proof since sets of the form $j^{-1} \circ \exp_G V$, V open and $V \subseteq W$, form an open basis at e in H. //

Remarks. (i) If $j : H \to G$ is a Lie subgroup of G, then

$$j \circ \exp_H = \exp_G \circ j_{*,e}.$$

It was mentioned after the proof of 5.1.2 that $\underline{\underline{h}}$ will usually be identified with its image in $\underline{\underline{g}}$ under the Lie algebra monomorphism $j_{*,e}$. Similarly we will usually suppress direct mention of the analytic monomorphism j. Thus

$$\exp_H = \exp_G|_{\underline{\underline{h}}}$$

and so, when no confusion is likely, both $\exp_G$ and $\exp_H$ will be denoted by exp. The remaining results and ideas of this section are of basic importance to the structure theorems of the next chapter and any reader who does not feel at ease with them is invited to fill in the missing monomorphisms.

(ii) Suppose that $j : H \to G$ is a Lie subgroup of G; the following two useful facts are easy to prove.

(a) If $\exp_G$ is one-to-one on a neighbourhood U of 0 in $\underline{\underline{g}}$, then $\exp_H$ is one-to-one on $U \cap \underline{\underline{h}}$ (or more accurately, $\exp_H$ is one-to-one on $j_{*,e}^{-1}(U \cap j_{*,e}(\underline{\underline{h}}))$).

(b) If U is as in (a) and $x \in H \cap U$, then there exists a unique $X \in U$ such that $\exp_G X = x$ and furthermore, $X \in \underline{\underline{h}}$. (More accurately, if U is as in (a) and $x \in j(H) \cap U$, then there exists a unique $X \in U$ such that $\exp_G X = x$ and furthermore, $X \in j_{*,e}(\underline{\underline{h}})$.)

5.2 Normal subgroups and ideals

The aim is to now take Theorem 5.1.3 a step further and show that the correspondence defined therein reduces to a correspondence defined between normal connected Lie subgroups of G and ideals of $\underline{\underline{g}}$. Before proving this result, further properties of the adjoint representation introduced in 4.2.2 are needed.

5.2.1 **Definitions.** (i) A subgroup H of a group G is said to be normal if $xHx^{-1} \subseteq H$ for all $x \in G$.

(ii) A Lie subalgebra L' of a Lie algebra L is said to be an ideal (of L) if $[X, Y] \in L'$ for all $X \in L$ and $Y \in L'$.

5.2.2 **Adjoint representation for Lie algebras.** Let L denote a Lie algebra. The map $ad : L \to \underline{\underline{gl}}(L)$ defined by

$$ad(X)(Y) = [X, Y], \text{ where } X, Y \in L,$$

is called the adjoint representation of L. Clearly ad is linear. It also preserves the Lie product. (As noted in §2.4, $\underline{\underline{gl}}(L)$ can be equipped with the structure of a Lie algebra in a natural manner; the Lie product of $M, N \in \underline{\underline{gl}}(L)$ is defined as $[M, N] = MN - NM$.) For if $X, Y, Z \in L$, the Jacobi identity shows that

$$\begin{aligned} ad[X, Y](Z) &= [[X, Y], Z] \\ &= -[[Y, Z], X] - [[Z, X], Y] \\ &= [X, [Y, Z]] - [Y, [X, Z]] \\ &= [ad\, X, ad\, Y](Z). \end{aligned}$$

Thus $ad : L \to \underline{\underline{gl}}(L)$ is a Lie homomorphism.

Returning to the case of a Lie group G with Lie algebra $\underline{\underline{g}}$, the adjoint representation $Ad : G \to GL(\underline{\underline{g}})$ of G is an analytic homomorphism when $GL(\underline{\underline{g}})$ is thought of as a Lie group (see §2.4). Thus $(Ad)_{*,e}$ is a Lie homomorphism from $\underline{\underline{g}}$ into $\underline{\underline{gl}}(\underline{\underline{g}})$. As we have just seen, the adjoint representation ad of $\underline{\underline{g}}$ is also a Lie homomorphism from $\underline{\underline{g}}$ into $\underline{\underline{gl}}(\underline{\underline{g}})$ ... the next result shows that these two homomorphisms are identical.

5.2.3 Theorem. Let G be a Lie group, Ad the adjoint representation of G, and ad the adjoint representation of $\underline{g}$. Then $(\mathrm{Ad})_{*,e} = \mathrm{ad}$.

Proof. Corollary 3.2.8 (i) of the BCH formula states that

$$(5.2.1) \quad \exp(tX)\exp(tY)\exp(-tX) = \exp\{tY + t^2[X, Y] + 0(t^3)\}$$

whenever $X, Y \in \underline{g}$. Using the definition of Ad and Exercise 2.F results in

$$A_{\exp tX} \circ \exp = \exp \circ \mathrm{Ad}(\exp tX).$$

Combining this formula with (5.2.1) yields

$$\exp\{\mathrm{Ad}(\exp tX)tY\} = \exp\{tY + t^2[X, Y] + 0(t^3)\}$$

and hence

$$(5.2.2) \quad \mathrm{Ad}(\exp tX)Y = Y + t[X, Y] + 0(t^2).$$

Select $X \in \underline{g}$ and define a 1-parameter subgroup ψ of G by $\psi : t \mapsto \exp tX$, and a 1-parameter subgroup θ of $GL(\underline{g})$ by $\theta : t \mapsto \mathrm{Ad}(\exp tX)$. From Theorem 2.2.10 there exists $X_0 \in \underline{\underline{gl}}(\underline{g})$ such that

$$\theta(t) = \exp tX_o \text{ and } X_0 = \theta_{*,e}[i, 1]_o.$$

Thus

$$\begin{aligned} X_0 &= \theta_{*,o}[i, 1]_o = (\mathrm{Ad} \circ \psi)_{*,o}[i, 1]_o \\ &= (\mathrm{Ad})_{*,e}\psi_{*,o}[i, 1]_o = (\mathrm{Ad})_{*,e}X. \end{aligned}$$

The proof is completed by showing that $X_0 = \mathrm{ad}(X)$. Let $Y \in \underline{g}$; then, using (5.2.2),

$$\begin{aligned} X_0(Y) &= \theta_{*,o}[i, 1](Y) = \frac{d}{dt}\{\mathrm{Ad}(\exp tX)\}\Big|_{t=o}(Y) \\ &= \frac{d}{dt}\{\mathrm{Ad}(\exp tX)(Y)\}\Big|_{t=o} \\ &= \frac{d}{dt}\{Y + t[X, Y] + 0(t^2)\}\Big|_{t=o} \end{aligned}$$

$$= [X,\ Y] = \mathrm{ad}(X)Y,$$

as required. (A word of explanation may be in order here. If f is a function from $\mathbf{R}$ into some linear space E, then in place of $f'(0) = f'(0)(1)$ it is often more convenient to write $\frac{d}{dt}\{f(t)\}\big|_{t=o}$.) //

5.2.4 **Remarks.** In section 2.4 it was shown that for the case of the general linear group $GL(n, \mathbf{R})$,

$$\exp X = e^X = I + X + \frac{X^2}{2!} + \dots$$

for $X \in \mathfrak{gl}(n, \mathbf{R})$, the Lie algebra of $GL(n, \mathbf{R})$. Thus, as a consequence of 5.2.3 and Exercise 2.F,

$$\mathrm{Ad}(\exp X) = e^{\mathrm{ad}\,X} = I + \mathrm{ad}\,X + \frac{(\mathrm{ad}\,X)^2}{2!} + \dots$$

for X in $\mathfrak{g}$.

The relationships between A_x, Ad and ad can be expressed concisely by the formulae

$$(A_x)_{*,e} = \mathrm{Ad}(x), \quad (\mathrm{Ad})_{*,e} = \mathrm{ad}$$

and the commuting diagram:

$$\begin{array}{ccccc} G & \xrightarrow{A_x} & G & \xrightarrow{\mathrm{Ad}} & GL(\mathfrak{g}) \\ \exp \uparrow & & \exp \uparrow & & \uparrow e^{(\cdot)} \\ \mathfrak{g} & \xrightarrow[\mathrm{Ad}(x)]{} & \mathfrak{g} & \xrightarrow[\mathrm{ad}]{} & \mathfrak{gl}(\mathfrak{g}) \end{array}$$

5.2.5 **Lemma.** *Consider the following conditions on a linear subspace* L *of* $\mathfrak{g}$, *where* $\mathfrak{g}$ *is the Lie algebra of a Lie group* G:

(i) $\mathrm{Ad}(x)L \subseteq L$ *for all* x *in* G (*and so* $\mathrm{Ad} : G \to GL(L)$ *is a well-defined representation of* G);

(ii) $\mathrm{ad}(X)L \subseteq L$ *for all* X *in* $\mathfrak{g}$ (*and so* $\mathrm{ad} : \mathfrak{g} \to \mathfrak{gl}(L)$ *is a well-defined representation of* $\mathfrak{g}$); *and*

(iii) L *is an ideal of* $\mathfrak{g}$.

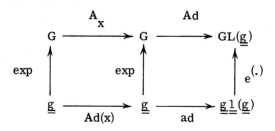

Then conditions (ii) and (iii) are equivalent and are implied by condition (i), while if G is also connected then all three conditions are equivalent.

Proof. Clearly conditions (ii) and (iii) are equivalent since $\mathrm{ad}(X)Y = [X, Y]$. Suppose that (i) is valid and let $X \in \underline{\underline{g}}$, $Y \in L$. Then

$$\begin{aligned}[X, Y] &= \mathrm{ad}(X)Y = (\mathrm{Ad})_{*, e}(X)Y \\ &= \frac{d}{dt}\{\exp t(\mathrm{Ad})_{*, e}(X)Y\}\Big|_{t=o} \\ &= \frac{d}{dt}\{\mathrm{Ad}(\exp tX)(Y)\}\Big|_{t=o}.\end{aligned}$$

However $\exp tX \in G$ and hence $\mathrm{Ad}(\exp tX)(Y) \in L$ for all $t \in \mathbf{R}$. Thus $\frac{d}{dt}\{\mathrm{Ad}(\exp tX)(Y)\}\Big|_{t=o} \in L$ showing that L is an ideal.

Now suppose that (ii) is valid along with the condition that G is connected. Let $X \in \underline{\underline{g}}$, $Y \in L$. Then

$$\begin{aligned}\mathrm{Ad}(\exp X)(Y) &= e^{\mathrm{ad}\,X}(Y) \\ &= Y + [X, Y] + \frac{1}{2!}[X, [X, Y]] + \dots .\end{aligned}$$

Since L is an ideal, each of the terms in the final expression belongs to L and so the sum converges to an element in L. The set $\exp(\underline{\underline{g}})$ is a neighbourhood of the identity of the connected group G and so 2.1.5 asserts that each element x in G is expressible as $\exp X_1 \dots \exp X_m$, where $X_1, \dots, X_m \in \underline{\underline{g}}$. Thus

$$\begin{aligned}\mathrm{Ad}(x)Y &= \mathrm{Ad}(\exp X_1 \dots \exp X_m)Y \\ &= \mathrm{Ad}(\exp X_1) \dots \mathrm{Ad}(\exp X_m)Y \in L,\end{aligned}$$

since $\mathrm{Ad}(\exp X_j)(L) \subseteq L$ for $j = 1, \dots, m$. //

5.2.6 **Theorem.** Let G be a connected Lie group. Then $H \mapsto \underline{\underline{h}}$ provides a one-to-one correspondence between all the normal connected Lie subgroups H of G and all the ideals of $\underline{\underline{g}}$.

Proof. As in Theorem 5.1.3, let L be a Lie subalgebra of $\underline{\underline{g}}$ and let H be the unique connected Lie subgroup of G with Lie algebra $\underline{\underline{h}} = L$. Suppose also that H is normal, that $X \in \underline{\underline{g}}$, and that $Y \in \underline{\underline{h}} = L$.

Then, just as in the first part of the proof of 5.2.5,

$$[X, Y] = \frac{d}{dt}\{Ad(\exp tX)Y\}\Big|_{t=o}.$$

However $\exp(Ad(\exp tX)Y) = \exp(tX).\exp(Y).\exp(-tX)$ and so $\exp(Ad(\exp tX)Y) \in H$ since $\exp(\pm tX) \in G$ for all $t \in \mathbf{R}$, $\exp Y \in H$, and H is normal. Hence $Ad(\exp tX)Y \in L$ for all $t \in \mathbf{R}$ by Lemma 5.1.4, implying that

$$[X, Y] = \frac{d}{dt}\{Ad(\exp tX)Y\}\Big|_{t=o} \in L,$$

as required.

On the other hand, suppose that $L = \underline{\underline{h}}$ is an ideal of $\underline{\underline{g}}$. Let $x \in G$, $X \in \underline{\underline{h}}$. From 5.2.5, $Ad(x)X \in L$ and thus

$$x.\exp(X).x^{-1} = A_x(\exp X) = \exp(Ad(x)X) \in \exp L.$$

Now use the fact that H is generated by $\exp L$ to show that $x.H.x^{-1} \subseteq H$ for all x in G and hence that H is normal. //

We will now take the preceding theorem a step further and describe the Lie subgroup of a Lie group G which has the centre of $\underline{\underline{g}}$ as its Lie algebra.

5.2.7 **Definitions.** (i) The <u>centre</u> of a Lie algebra L is defined as the ideal $\{X \in L : [X, Y] = 0$ for all $Y \in L\}$.

(ii) The <u>centre</u> of a group G is defined as the normal subgroup $\{x \in G : xy = yx$ for all $y \in G\}$.

Clearly the two definitions are analogous since the centre of L can also be expressed as $\{X \in L : [X, Y] = [Y, X]$ for all $Y \in L\}$.

5.2.8 **Corollary.** <u>The identity component</u> J <u>of the centre of a connected Lie group</u> G <u>is a closed Abelian Lie subgroup of</u> G <u>and its Lie algebra</u> $\underline{\underline{j}}$ <u>is the centre of</u> $\underline{\underline{g}}$. <u>Furthermore,</u>

$$\underline{\underline{j}} = \{X \in \underline{\underline{g}} : Ad(x)X = X \text{ for all } x \text{ in } G\}.$$

5.2.9 **Remark.** Even for the case of a connected Lie group it may happen that its centre is not connected. For example, the centre of the compact connected Lie group SU(2) is $\{I, -I\}$, where I is the 2×2 identity matrix.

Proof (of 5. 2. 8). Put $L = \{X \in \underline{\underline{g}} : Ad(x)X = X \text{ for all } x \text{ in } G\}$. We will first show that L is equal to the centre of $\underline{\underline{g}}$. If $X \in L$ and $Y \in \underline{\underline{g}}$ we can use the first paragraph of the proof of 5. 2. 5 to deduce that X belongs to the centre of $\underline{\underline{g}}$ since

$$[Y, X] = \frac{d}{dt}\{Ad(\exp tY)X\}\Big|_{t=o}$$
$$= \frac{d}{dt}\{X\}\Big|_{t=o} = 0.$$

On the other hand, if X belongs to the centre of $\underline{\underline{g}}$ and if $Y \in \underline{\underline{g}}$, then

$$Ad(\exp Y)X = e^{ad\,Y}(X)$$
$$= X + [Y, X] + \frac{1}{2!}[Y, [Y, X]] + \dots$$
$$= X.$$

Since G is connected it is generated by $\exp(\underline{\underline{g}})$ and hence $Ad(x)X = X$ for all x in G. Thus $X \in L$ and so L is equal to the centre of $\underline{\underline{g}}$ as asserted.

We will now demonstrate that L is the Lie algebra of J, the identity component of the centre of G. Let J_o denote the unique Abelian connected normal Lie subgroup of G which has L as its Lie algebra (Theorem 5. 2. 6). If $X \in L$ and $x \in G$, then

$$x. \exp X. x^{-1} = \exp(Ad(x)X) = \exp X.$$

However, J_o is generated by $\exp(L)$ and so J_o is contained in the centre of G. Thus $J_o \subseteq J$. Since J is closed in the centre of G and the centre of G is closed in G, then J is closed in G. Also J is connected and Abelian, and so $j : J \to G$ is a (closed) connected Abelian Lie subgroup of G (Theorem 3. 3. 1). Since $J_o \subseteq J$, then $L \subseteq \underline{\underline{j}}$, where $\underline{\underline{j}}$ is the Lie algebra of J.

Let $X \in \underline{\underline{j}}$ and $x \in G$. By Lemma 5. 1. 4, $\exp X \in J$ and hence

$$\exp(Ad(x)X) = x. \exp X. x^{-1} = \exp X.$$

By first considering a real multiple of X if necessary, we see that $\mathrm{Ad}(x)X = X$ for all $x \in G$ and $X \in \underline{\underline{j}}$. Thus $\underline{\underline{j}} \subseteq L$, whence the required $\underline{\underline{j}} = L$. //

5.2.10 **Lie algebras of products.** As a further application of 5.2.6 we will combine it with the CBH formula to give a description of the Lie algebra of the product of a finite number of connected Lie groups. Suppose that G_1 and G_2 are connected Lie groups with Lie algebras $\underline{\underline{g}}_1$ and $\underline{\underline{g}}_2$ respectively, and that the Lie algebra of $G = G_1 \times G_2$ is $\underline{\underline{g}}$. (This group G is indeed a Lie group by 2.2.16.) Since G_1 (or more accurately, $G_1 \times \{e_2\}$) is a closed normal subgroup of G, then $\underline{\underline{g}}_1$ is an ideal of $\underline{\underline{g}}$. Similarly $\underline{\underline{g}}_2$ is also an ideal of $\underline{\underline{g}}$.

If $X \in \underline{\underline{g}}_1 \cap \underline{\underline{g}}_2$, then $\exp tX \in G_1 \cap G_2 = \{e\}$ for all real t and so $\underline{\underline{g}}_1 \cap \underline{\underline{g}}_2 = \{0\}$. Furthermore

$$\exp X_1 \,.\, \exp X_2 = \exp X_2 \,.\, \exp X_1$$

(in other words,

$$(\exp_{G_1} X_1, e_2)(e_1, \exp_{G_2} X_2) = (e_1, \exp_{G_2} X_2)(\exp_{G_1} X_1, e_2)),$$

and so we may deduce from thw CBH formula that

$$\exp X_1 \,.\, \exp X_2 = \exp(X_1 + X_2).$$

The dimension of $\underline{\underline{g}}$ must be equal to the sums of the dimensions of $\underline{\underline{g}}_1$ and $\underline{\underline{g}}_2$ and hence $\underline{\underline{g}} = \underline{\underline{g}}_1 \oplus \underline{\underline{g}}_2$. Clearly the arguments used here can be extended to a finite product. Also the identity component of a product of topological groups is equal to the product of their identity components (Exercise), and the Lie algebra of the (closed) identity component of a Lie group is equal to the Lie algebra of the whole group (use Lemma 2.1.4). Thus:

Proposition. Let $G_1, \ldots, G_m$ be a family of m Lie groups with Lie algebras $\underline{\underline{g}}_1, \ldots, \underline{\underline{g}}_m$ respectively. If $\underline{\underline{g}}$ is the Lie algebra of $G = G_1 \times \ldots \times G_m$, then each $\underline{\underline{g}}_j$ $(j = 1, \ldots, m)$ is an ideal of $\underline{\underline{g}}$ and

$$\underline{\underline{g}} = \underline{\underline{g}}_1 \oplus \ldots \oplus \underline{\underline{g}}_m .$$

Furthermore, if $X_j \in \underline{\underline{g}}_j$ $(j = 1, \ldots, m)$, then

$$\exp_G(X_1 + \ldots + X_m) = \exp_G X_1 \ldots \exp_G X_m = (\exp_{G_1} X_1, \ldots, \exp_{G_m} X_m).$$

Notes

In the next chapter we will see how to decompose certain Lie algebras (in particular, Lie algebras of compact groups) into direct sums of their centres and all their simple ideals. Accordingly, when we have the Lie algebra of a compact group, the above correspondence results show that such a decomposition can be transported to a similar decomposition of the compact group.

Forerunners of the results of this chapter can be found in Lie's original papers on transformation groups. For example, given a Lie transformation group G of the plane, Lie examined certain types of subalgebras of its Lie algebra (of infinitesimal transformations) and then described the subgroups of G which have these subalgebras as their Lie algebras. See Section 14.5 of Lie [1].

Exercises

5.A. Let G, H be Lie groups and $\phi : H \to G$ an analytic homomorphism. Show that ϕ is an immersion if and only if $\phi_{*,e}$ is an injection.

5.B. Show that the set of points $(e^{it}, e^{i\sqrt{2}t})$, with $t \in \mathbf{R}$, when equipped with the topology described in the proof of Theorem 5.1.3, forms a noncompact Lie subgroup of $\mathbf{T}^2$. Show also that this topology is strictly stronger than that induced from $\mathbf{T}^2$. (Compare with Exercise 2.B and Theorem 6.3.13.) Observe that when this 1-parameter subgroup of $\mathbf{T}^2$ is equipped with the induced topology it is not a Lie group since it is not even locally compact!

5.C. Let G be a Lie group with Lie algebra $\underline{\underline{g}}$ and let $\underline{\underline{h}}_1$ and $\underline{\underline{h}}_2$ be the Lie algebras of connected Lie subgroups H_1 and H_2 of G. Then $\underline{\underline{h}}_1$ and $\underline{\underline{h}}_2$ are subalgebras of $\underline{\underline{g}}$.

(i) Show that the connected Lie subgroup of G corresponding to the Lie subalgebra $\underline{\underline{h}}_1 \cap \underline{\underline{h}}_2$ is algebraically equal to $H_1 \cap H_2$.

(ii) Let H denote the subgroup of G generated by H_1 and H_2. Show that the connected Lie subgroup of G corresponding to the Lie subalgebra of $\underline{\underline{g}}$ generated by $\underline{\underline{h}}_1$ and $\underline{\underline{h}}_2$ lies between H and its closure in G.

6·Characterisations and structure of compact Lie groups

The first section of this chapter is concerned with establishing a number of conditions which are equivalent to a compact group being Lie. These include the conditions that G has no small subgroups and that G is isomorphic to a closed subgroup of some unitary group. Thus all compact groups are linear; section 2 contains an example due to Birkhoff of a connected Lie group which is not linear. Before providing the structure theorems of section 4 it is necessary (in section 3) to analyse simple and semisimple Lie algebras and Lie groups. In the main this will be done using a powerful tool, the Killing form of a Lie algebra. Such an analysis allows a proof of the fact that all semisimple connected Lie subgroups of a compact Lie group are closed. This result finally completes the mise en scène for the entrance of the promised structure theorems; section 4 contains these theorems for compact connected Lie groups while section 5 contains them for arbitrary compact connected groups.

6.1 Compact groups and Lie groups

The main result in this section is a list of necessary and sufficient conditions for a compact group to be a Lie group. As explained in Appendix A, the dual hypergroup Γ of a compact group G is a maximal set of pairwise inequivalent, continuous, unitary, irreducible representations of G. Whenever $\Delta \subseteq \Gamma$, then $[\Delta]$ is the subset of Γ generated by the elements in Δ as described in A.3.3. If $\Gamma = [\Delta]$ where Δ is finite, then Γ is said to be finitely-generated.

6.1.1 **Theorem.** The following conditions on a compact group G are equivalent:

(i) G is a Lie group;

(ii) G has no small subgroups;

(iii) there exists a topological isomorphism from G onto a closed subgroup of some $O(m)$;

(iv) there exists a topological isomorphism from G onto a closed subgroup of some $U(n)$;

(iv) the dual hypergroup of G is finitely-generated.

The following lemma is used in the proof of the above result.

6.1.2 **Lemma.** Let G be a compact group and let U be a neighbourhood of e in G. There exists a continuous homomorphism β from G into some orthogonal group $O(m)$ such that $\ker(\beta) \subseteq U$.

Proof. Put $V = U \cap U^{-1}$; then V is a symmetric neighbourhood of e in G. Choose f in $C(G, \mathbf{R})$ to satisfy $f(e) > 0$ and $\text{supp}(f) \subseteq V$. (This choice is possible by Theorem (8.4) of Hewitt and Ross [1].) From A.3.6 of the Appendix we know that f can be uniformly approximated by trigonometric polynomials and so there exists a trigonometric polynomial g such that $\text{re}(g(x)) < \text{re}(g(e))$ for x in $G\backslash V$. Define $h = \frac{1}{2}(g + \overline{g})$; then h is a real-valued trigonometric polynomial which satisfies

$$(6.1.1) \quad h(x) < h(e) \text{ for } x \text{ in } G\backslash V.$$

Let F denote the smallest closed subspace of $C(G, \mathbf{R})$ which contains h and all its left translates. (The left translate of h by a is defined as the function $\tau_a h : x \mapsto h(a^{-1}x)$.) Then F is finite-dimensional (why?) of dimension m, say. Define a continuous homomorphism β from G into $GL(F)$ by

$$\beta : x \mapsto (\tau_x : F \to F).$$

Whenever $x \in G\backslash U$, then $x^{-1} \in G\backslash V$ and so $(\tau_x h)(e) = h(x^{-1}) < h(e)$ from (6.1.1), showing that x is not a member of $\ker(\beta)$. Thus $\ker(\beta) \subseteq U$. If F is now given the inner product induced from $L^2(G, \mathbf{R})$, then β becomes an orthogonal representation. Select a real orthonormal basis for F of dimension m and identify each $\beta(x)$ with a member of $O(m)$ in the usual manner. //

Proof (of 6.1.1). The proof is little more than a piecing together of results already obtained. The actual implications proved are represented by:

$$\text{(i)} \Rightarrow \text{(ii)} \Rightarrow \text{(iii)} \Rightarrow \text{(iv)} \Rightarrow \text{(i)}, \quad \text{(iv)} \Longleftrightarrow \text{(v)}.$$

The fact that (i) implies (ii) follows directly from Corollary 2.2.14. Assume that there exists a neighbourhood U of e in G which doesn't contain any nontrivial subgroups of G (that is, G has no small subgroups). From the preceding lemma there exists a continuous homomorphism β from G into some orthogonal group $O(m)$ such that $\ker(\beta) \subseteq U$. But $\ker(\beta)$ is a subgroup of G and so $\ker(\beta) = \{e\}$. Thus β is an injection. Since G is compact so is $\beta(G)$. Also β^{-1} is continuous on $\beta(G)$ for if $\beta(x_\lambda) \to \beta(x)$ in $\beta(G)$, then x_λ is a net in the compact space G and so must have a limit point; this limit point can be none other than x. Thus β is a topological isomorphism onto a closed subgroup of $O(m)$.

The implication '(iii) implies (iv)' is trivial since $O(m)$ is a closed subgroup of $U(m)$. Now suppose that (iv) is satisfied, that is that we have a topological isomorphism β from G onto a closed subgroup H of some $U(m)$. Theorem 3.3.1 shows that H is a Lie group and since it is true that all topological isomorphs of Lie groups are Lie groups (Exercise), it follows that G is also a Lie group.

To see that (iv) implies (v), let β denote a topological isomorphism from G onto a closed subgroup of $U(n)$. Then β is a continuous unitary representation of G and so can be decomposed in the form $\beta = \gamma_1 \oplus \dots \oplus \gamma_k$, where each γ_j is a member of Γ, the dual hypergroup of G (see A.2.3 (ii)). Furthermore, the faithfulness of β ensures that $[\gamma_1, \dots, \gamma_k] = \Gamma$ (see A.3.4). On the other hand, suppose that there exists $\gamma_1, \dots, \gamma_k$ in Γ which generate Γ. Then the continuous unitary representation

$$x \mapsto \gamma_1(x) \oplus \dots \oplus \gamma_k(x)$$

is faithful showing that G is topologically isomorphic to a compact, hence closed, subgroup of $U(d(\gamma_1) + \dots + d(\gamma_k))$. Thus (v) implies (iv). //

6.1.3 **Corollaries.** The following are some simple but important consequences of Theorem 6.1.1.

(a) All compact Abelian Lie groups are of the form $T^r \times F$, where T is the circle group $(= U(1))$, r is a non-negative integer, and F is a finite Abelian group. (This is immediate from the fact that if G is compact Abelian, then Γ is a discrete Abelian group, and if it is also finitely generated, then it is well-known that it is of the form $Z^r \times F$, where Z is the additive group of integers and F is a finite Abelian group (Hewitt and Ross [1, (A.27)]).) This description of compact connected Abelian Lie groups answers Exercise 3.B (iv) while 3.B (iii) calls for a proof that a group is a connected Abelian Lie group if and only if it is isomorphic to $T^m \times R^k$, where m and k are integers.

(b) Let G be a topological group and H a closed normal subgroup. Then G/H is a (Hausdorff) topological group when equipped with its usual topology, namely the strongest topology which allows the natural projection $\pi : G \to G/H$ to be continuous. Since π is also open (Hewitt and Ross [1, (5.26)]) G/H has no small subgroups whenever G has this property, whence G/H is a compact Lie group whenever G is a compact Lie group.

(c) Let G be a topological group and H a subgroup of G. If H and G/H are compact Lie groups, then G itself is compact Lie. (Theorem (5.25) of Hewitt and Ross [1] shows that G must be compact. Since H and G/H are compact Lie groups they have no small subgroups; we will show that the same is true for G. Let U be a neighbourhood of the identity in G/H containing no nontrivial subgroups and let V denote its inverse image under π in G. Let W denote a neighbourhood of the identity in G such that $W \cap H$ contains no nontrivial subgroups. Then $W \cap V$ is a neighbourhood of the identity in G which contains no subgroups apart from $\{e\}$. For if K is a subgroup in $W \cap V$, then $\pi(K)$ is a subgroup of U so that K must be a subset of H. In this case K is a subgroup of H contained $W \cap H$ whence K can be nothing but the trivial group.)

Results (b) and (c) above are known to hold for general, not necessarily compact, Lie groups and can be proved directly by constructing the appropriate atlases. On the other hand the simple arguments used here are immediately applicable to the general case if one first proves the very deep result due to Yamabe that a locally compact group is a Lie group if (and only if) it has no small subgroups. See the notes at the end of this chapter for further details.

6.2 Linear Lie groups

The conclusion of Theorem 6.1.1 leads us to ask whether or not every Lie group is topologically isomorphic with some group of real or complex matrices. In this section we will show that the answer is 'no'.

6.2.1 Definition. Let G be a Lie group which is a topological subgroup of GL(n, **R**) for some $n > 0$; then G, and all Lie groups topologically isomorphic with it, are said to be real linear Lie groups of degree n. Complex linear Lie groups are defined in an analogous manner with GL(n, **C**) in place of GL(n, **R**).

Notice that all the examples of Lie groups given in Chapter 2 are linear Lie groups. Also Theorem 6.1.1 shows that every compact Lie group is a real linear Lie group. In fact, it wasn't until 1935 that the first example of a nonlinear connected Lie group was given. This example was due to Birkhoff [1]. (Exercise 6.C describes a method of constructing a discrete Abelian group (hence a 0-dimensional Lie group) which is not a linear Lie group. From this it is a simple matter to construct Lie groups of any prescribed dimension which are not linear.) The dimension of Birkhoff's example is 3 and this is best possible since:

Fact. The only connected Lie groups (up to isomorphism) of dimension 1 are **T** and **R** and of dimension 2 are $\mathbf{T}^k \times \mathbf{R}^{2-k}$, $k = 0, 1, 2$, and $GA(\mathbf{R})^+$. (For the definition of $GA(\mathbf{R})^+$ see Exercise 2.I.)

Proof. To verify the 1-dimensional case is the goal of Exercise 3.F. For the 2-dimensional case, first note that by Exercise 3.B the only 2-dimensional connected Abelian Lie groups are $\mathbf{T}^k \times \mathbf{R}^{2-k}$, $k = 0, 1, 2$. Now $GA(\mathbf{R})^+$ is a 2-dimensional simply-connected non-Abelian Lie group and from Exercise 2.I and the remarks on covering groups in 3.4.7, it is the unique example of such groups. Let G be a connected 2-dimensional non-Abelian Lie group and let $p : GA(R)^+ \to G$ be a covering of G. Since p is homomorphism which is both continuous and open, G is isomorphic to $GA(R)^+/p^{-1}(e)$. But p is also a local isomorphism and so $p^{-1}(e)$ must be discrete, hence central by Lemma 6.4.1 below. But $GA(R)^+$ has a trivial centre. //

6.2.2 **Lemma.** Let G be any subgroup of GL(n, C). Suppose that G contains elements S and T whose commutator $R = S^{-1}T^{-1}ST$ is of prime order p, $p > 1$, and satisfies $SR = RS$, $TR = RT$. Then $n \geq p$.

Proof. Since $R \in GL(n, C)$ there exists a nonzero $\alpha \in C$ and a nonzero $x_0 \in C^n$ such that $Rx_0 = \alpha x_0$. Furthermore α can be chosen to be a primitive p-th root of unity (Exercise 6.A). Now let $\mathcal{H}$ denote the linear subspace of all vectors x in C^n such that $Rx = \alpha x$. If $x \in \mathcal{H}$, then

$$R(Sx) = S(Rx) = S(\alpha x) = \alpha(Sx)$$

showing that $\mathcal{H}$ is closed under the action of S, and likewise of T.

Considering S and T as operators on $\mathcal{H}$, $R = S^{-1}T^{-1}ST = \alpha I$ (I denoting the identity operator on $\mathcal{H}$) whence $T^{-1}ST = \alpha S$. A standard result in the theory of operators on finite-dimensional spaces shows that $T^{-1}ST$ and S have the same eigenvalues, and hence so do S and αS. Since S is nonsingular it has a nonzero eigenvalue, β say. Suppose that $Sy = y$, where y is a nonzero vector in $\mathcal{H}$. Then

$$\alpha Sy = \alpha\beta y$$

and so $\alpha\beta$ is an eigenvalue of αS. But we have just seen that if this is the case, then it must also be an eigenvalue of S. Continuing in this fashion yields the conclusion that S has at least p eigenvalues, namely $\beta, \alpha\beta, \alpha^2\beta, \ldots, \alpha^{p-1}\beta$. From this it follows that $n \geq p$. //

Let G_3 denote the group of matrices

$$M(x, y, z) = \begin{pmatrix} 1 & x & z \\ 0 & 1 & y \\ 0 & 0 & 1 \end{pmatrix}, \text{ where } x, y, z \in \mathbf{R},$$

and let N denote the discrete normal subgroup of G_3 consisting of the matrices $M(0, 0, n)$, $n \in \mathbf{Z}$. Then $G_3^* = G_3/N$ is a connected Lie group of dimension 3. (This is Exercise 6.B. Roughly speaking, G_3^* consists of elements $M(x, y, z)$, where $x, y \in \mathbf{R}$ and $z \in \mathbf{R}/\mathbf{Z}$.)

6.2.3 **Proposition.** The group G_3^* is a connected Lie group of dimension 3 which is neither a real nor complex Lie group.

Proof. Suppose that G_3^* is isomorphic with a subgroup of $GL(n, C)$. Let π denote the canonical projection from G_3 onto G_3^*. Let p denote a prime number; then the images S, T of $\pi M(1, 0, 0)$ and $\pi M(0, 1/p, 0)$ in $GL(n, C)$ both commute with $R = S^{-1}T^{-1}ST$, which is the image of $\pi M(0, 0, 1/p)$. Now R is of order p so that from the above lemma, $n \geq p$. Since p was chosen arbitrarily, we have a contradiction. //

6.3 Simple and semisimple Lie algebras

For our purposes the main result of this section is that if L is a semisimple Lie subalgebra of $\underline{g}$, the Lie algebra of a compact group G, then the connected Lie subgroup of G corresponding to L is closed in G. For the precise role played by this result in the first structure theorem of the next section see the Notes at the end of the chapter.

6.3.1 **Definitions.** (i) A Lie algebra is simple [resp. semisimple] if it is noncommutative and has no proper nontrivial ideals [resp. no proper nontrivial commutative ideals].

(ii) A Lie group is simple [resp. semisimple] if its Lie algebra is simple [resp. semisimple].

Hence, by Exercise 3.A and Theorem 5.2.6, a connected Lie group is simple if and only if it is non-Abelian and has no proper, nontrivial, connected, normal subgroups. In 6.4.3 below we will see that a compact connected Lie group is semisimple if and only if it has a finite centre.

6.3.2 **Lemma.** (i) Every simple Lie algebra is semisimple.

(ii) A semisimple Lie algebra has centre $\{0\}$.

(iii) If L is a Lie algebra which has a decomposition

$$L = L_0 \oplus L_1 \oplus \ldots \oplus L_m,$$

where L_0 is the centre of L and L_j $(1 \leq j \leq m)$ are simple ideals in L, then the L_j $(1 \leq j \leq m)$ are all the simple ideals in L and so this decomposition is unique. Furthermore, the Lie algebra $L^{\#} = L_1 \oplus \ldots \oplus L_m$ is semisimple.

6.3.3 **Remark.** Before proceeding further, it is worth pointing out that as a corollary of the main structure theorem - Theorem 6.4.2 - the above lemma can be considerably strengthened for Lie algebras which are Lie algebras of compact groups. (Such Lie algebras are referred to as compact Lie algebras.) Suppose that L is the Lie algebra of a compact group; then it will be seen that L always has a (necessarily unique) decomposition as depicted in 6.3.2 (iii). As a consequence, L is semisimple when (and only when) it has a trivial centre. (See 6.4.3 below.) At the same time we see that a compact Lie algebra has a decomposition into simple ideals if and only if it is semisimple; however this result is true for all Lie algebras and follows easily from Cartan's criterion for semisimplicity (see 6.3.6 (ii)).

Proof (of 6.3.2). The proof of (i) is immediate. For the proof of (ii) it suffices to notice that the centre of a Lie algebra is a commutative ideal.

Suppose now that a Lie algebra L has a decomposition as described in (iii). Let K be a commutative ideal of $L^{\#} = L_1 \oplus \ldots \oplus L_m$. Then $[K, L_j] \subseteq K \cap L_j$ $(1 \leq j \leq m)$ and since L_j is simple, either $K \supseteq L_j$ or $K \cap L_j = \{0\}$. The former case is impossible since L_j is noncommutative and hence $[K, L_j] = \{0\}$ for $j = 1, \ldots, m$. Thus K is contained in the centre of $L^{\#}$. Let X belong to the centre of $L^{\#}$ and let $Y \in L$. Then $Y = Y_0 + Y^{\#}$, where $Y_0 \in L_0$ and $Y^{\#} \in L^{\#}$. Therefore

$$[X, Y] = [X, Y_0] + [X, Y^\#] = 0$$

(the first term is 0 because Y_0 belongs to L_0, the second because X belongs to the centre of $L^\#$), showing that X also belongs to L_0, the centre of L. Thus the centre of $L^\#$ must be trivial and in particular, K is trivial. Thus $L^\#$ is semisimple.

Let us now suppose that K is a simple ideal of L. For $1 \le j \le m$, $[K, L_j] \subseteq K \cap L_j$ and since $K \cap L_j$ is also a simple ideal of L, either $K = L_j$ for some j or $K \cap L_j = \{0\}$ for all j. If the latter, then K belongs to the (commutative) centre of $L^\#$ and hence $K = \{0\}$. //

Simple and semisimple Lie algebras are often approached via a particular bilinear form known as the Killing form.

6.3.4 **The Killing form.** Let L be a (real, finite-dimensional) Lie algebra; the bilinear form

$$K(X, Y) = \mathrm{tr}(\mathrm{ad}\, X\, \mathrm{ad}\, Y),$$

where tr denotes the usual trace (in this case the trace of the linear operator $\mathrm{ad}\, X\, \mathrm{ad}\, Y : L \to L$), is called the Killing form of L.

It follows without difficulty that the restriction of the Killing form to a Lie ideal is the Killing form of the ideal. Also, since $\mathrm{tr}(AB) = \mathrm{tr}(BA)$ for endomorphisms A, B, we have that the Killing form is symmetric. This fact combines with the Jacobi identity to furnish

$$\begin{aligned} \mathrm{tr}(\mathrm{ad}[X, Y]\mathrm{ad}\, Z) &= \mathrm{tr}((\mathrm{ad}\, X\, \mathrm{ad}\, Y - \mathrm{ad}\, Y\, \mathrm{ad}\, X)\mathrm{ad}\, Z) \\ &= \mathrm{tr}(\mathrm{ad}\, X\, \mathrm{ad}\, Y\, \mathrm{ad}\, Z) - \mathrm{tr}(\mathrm{ad}\, Y\, \mathrm{ad}\, X\, \mathrm{ad}\, Z) \\ &= \mathrm{tr}(\mathrm{ad}\, Y\, \mathrm{ad}\, Z\, \mathrm{ad}\, X) - \mathrm{tr}(\mathrm{ad}\, Z\, \mathrm{ad}\, Y\, \mathrm{ad}\, X) \\ &= \mathrm{tr}((\mathrm{ad}\, Y\, \mathrm{ad}\, Z - \mathrm{ad}\, Z\, \mathrm{ad}\, Y)\mathrm{ad}\, X) \\ &= \mathrm{tr}(\mathrm{ad}[Y, Z]\mathrm{ad}\, X), \end{aligned}$$

with the consequence that

$$(6.3.1) \quad K([X, Y], Z) = K([Y, Z], X) = K([Z, X], Y)$$

for all $X, Y, Z \in L$. (Any symmetric bilinear form on a Lie algebra which satisfies (6.3.1) is said to be invariant.)

The Killing form on a Lie algebra L is said to be nondegenerate if $K(X, Y) = 0$ for all $Y \in L$ occurs only when $X = 0$; otherwise K is said to be degenerate. For example, if L is commutative then $K(X, Y) = 0$ for all $X, Y \in L$. The other extreme is given by:

6.3.5 **Cartan's criterion for semisimplicity.** A Lie algebra is semisimple if and only if its Killing form is nondegenerate.

This result, along with a related criterion for a Lie algebra to be solvable (also due to É. Cartan), lies at the heart of the theory of Lie algebras. We will not prove it here, but refer the interested reader to any standard text on Lie algebras (for example, Jacobson [1, Chapter III] or Samelson [1, p. 25]).

Cartan's criterion for semisimplicity has many important corollaries, but only two of them will be presented here.

6.3.6 **Corollary.** (i) *If* L *is a semisimple Lie algebra and* M *is an ideal of* L, *then* M *is semisimple along with* $M^{\perp} = \{X \in L : K(X, Y) = 0$ *for all* $Y \in M\}$. *Also*

$$M \cap M^{\perp} = \{0\} \text{ and } M \oplus M^{\perp} = L.$$

(ii) *A Lie algebra* L *is semisimple if and only if it has a decomposition as a direct sum*

$$L = L_1 \oplus \dots \oplus L_m,$$

where $L_1, \dots, L_m$ *are simple ideals of* L.

Proof. Begin the proof of (i) by letting M be an ideal of semisimple L; in view of (6.3.1), $M^{\perp}$ is also an ideal. If $X, Y \in M \cap M^{\perp}$ and $Z \in L$, then

$$K([X, Y], Z) = K([Y, Z], X) = 0$$

since $[Y, Z] \in M^{\perp}$ and $X \in M$. Since the Killing form on L is nondegenerate we have that $M \cap M^{\perp}$ is a commutative ideal of L. But L is semisimple and so the only possibility is $M \cap M^{\perp} = \{0\}$.

Let $a_1, \ldots, a_m$ be a basis for M and define a linear operator $\phi : L \to M$ by

$$\phi(X) = K(a_1, X)a_1 + \ldots + K(a_m, X)a_m.$$

The kernel of ϕ is $M^\perp$ ($\phi(X) = 0$ iff $K(a_1, X) = \ldots = K(a_m, X) = 0$ iff $K(Y, X) = 0$ for all $Y \in M$) and hence, using the idea of Lemma 1.2.5,

$$\begin{aligned}\dim(L) &= \dim(\operatorname{Im} \phi) + \dim(\operatorname{Ker} \phi) \\ &\leq \dim(M) \quad + \dim(M^\perp).\end{aligned}$$

But the opposite inequality is also valid since

$$\dim(L) \geq \dim(M \oplus M^\perp) = \dim(M) + \dim(M^\perp),$$

and so it is a fact that $L = M \oplus M^\perp$. Combine this decomposition with Cartan's criterion and the fact that the Killing form K on M is the restriction of K to $M \times M$ to show that M is semisimple. Since M was chosen arbitrarily in L, then $M^\perp$ must also be semisimple. (Some of the steps of this proof are isolated in Exercise 6. D.) The proof of (ii) follows by finite induction. //

The second application of Cartan's criterion for semisimplicity involves the notion of a derivation.

6. 3. 7 **Definition.** Let L be a Lie algebra and let ϕ be a linear map from L into L. Then ϕ is said to be a derivation of L if

$$\phi[X, Y] = [\phi X, Y] + [X, \phi Y]$$

for $X, Y \in L$.

The set of all derivations of L is denoted by $\partial(L)$ ... some straightforward manipulations show that $\partial(L)$ is a Lie subalgebra of $\underline{\underline{gl}}(L)$. An example of a derivation is ad(X), where $X \in L$, and all such derivations are referred to as inner.

6.3.8 **Corollary.** *If the Killing form of a Lie algebra L is nondegenerate, then all derivations of L are inner. (Thus all derivations of a semisimple Lie algebra are inner.)*

Proof. Let D be a derivation of L and define a linear mapping $X \mapsto \mathrm{tr}(D \circ \mathrm{ad}\, X)$ from L into **C**. Since K is nondegenerate, there exists $X_0 \in L$ such that

$$\mathrm{tr}(D \circ \mathrm{ad}\, X) = K(X_0, X) \quad \text{for all } X \in L.$$

Let E denote a derivation of L. For X, Y $\in$ L we have

$$\begin{aligned} K(EX, Y) &= \mathrm{tr}(\mathrm{ad}\, EX \circ \mathrm{ad}\, Y) \\ &= \mathrm{tr}((E\, \mathrm{ad}\, X - \mathrm{ad}\, XE)\mathrm{ad}\, Y) \\ &= \mathrm{tr}(E\, \mathrm{ad}\, X\, \mathrm{ad}\, Y - E\, \mathrm{ad}\, Y\, \mathrm{ad}\, X) \\ &= \mathrm{tr}(E \circ [\mathrm{ad}\, X, \mathrm{ad}\, Y]) \\ &= \mathrm{tr}(E \circ \mathrm{ad}[X, Y]). \end{aligned}$$

Now suppose that $E = D - \mathrm{ad}\, X_0$ and put $Z = [X, Y]$. Then

$$\begin{aligned} \mathrm{tr}(E \circ \mathrm{ad}\, Z) &= \mathrm{tr}(D \circ \mathrm{ad}\, Z) - \mathrm{tr}(\mathrm{ad}\, X_0 \circ \mathrm{ad}\, Z) \\ &= K(X_0, Z) - K(X_0, Z) = 0. \end{aligned}$$

Since K is nondegenerate, and $K(EX, Y) = \mathrm{tr}(E \circ \mathrm{ad}[X, Y])$ for all X, Y $\in$ L, then $E = 0$ whence $D = \mathrm{ad}\, X_0$. //

6.3.9 **Group of automorphisms.** Let L denote a Lie algebra and let Aut(L) denote the group of Lie algebra automorphisms of L. Then Aut(L) is a closed subgroup of GL(L) and hence, by virtue of Theorem 3.3.1, is a Lie group. From §2.4 we know that its Lie algebra is a subalgebra of $\underline{\underline{gl}}(L)$ - we will now describe explicitly this Lie subalgebra. From the Remark following the proof of Theorem 3.3.1 (the promise there being fulfilled by Lemma 5.1.4), D belongs to the Lie algebra of Aut(L) if and only if $D \in \underline{\underline{gl}}(L)$ and $e^{tD} \in \mathrm{Aut}(L)$ for all $t \in \mathbf{R}$. Suppose that D does belong to the Lie algebra of Aut(L) and let X, Y $\in$ L. Then e^{tD} is an automorphism for each $t \in \mathbf{R}$ and so

$$e^{tD}[X, Y] = [e^{tD}X, e^{tD}Y],$$

whence $D[X, Y] = [DX, Y] + [X, DY]$. Thus D is a derivation of L.

Now suppose that D is a derivation of L. Induction on the formula $D[X, Y] = [DX, Y] + [X, DY]$ yields

$$D^k[X, Y] = \sum_{m+n=k} \frac{k!}{m!\,n!} [D^m X, D^n Y]$$

for integers $m, n \in \{0, 1, \dots\}$, where D^0 is interpreted as the identity automorphism of L. From this family of identities it follows that

$$e^{tD}[X, Y] = [e^{tD}X, e^{tD}Y]$$

for all $t \in \mathbf{R}$ and $X, Y \in L$. We know from the discussion of the general linear group in §2.4 that $e^{tD} \in GL(L)$ and hence $e^{tD} \in \mathrm{Aut}(L)$ for all $t \in \mathbf{R}$. Thus, whenever L is a Lie algebra, the Lie algebra of $\mathrm{Aut}(L)$ is precisely $\partial(L)$.

When L is semisimple this result can be made even more decisive.

6.3.10. **Corollary.** Let L be a semisimple Lie algebra. Then the Lie algebra of $\mathrm{Aut}(L)$ is (isomorphic to) L.

Proof. In view of the preceding remarks, the only requirement is a demonstration that L and $\partial(L)$ are isomorphic. Define a map $X \mapsto \mathrm{ad}\, X$ from L into $\partial(L)$. Clearly it is homomorphism and furthermore it is onto (Corollary 6.3.8). The kernel of ad is the centre of L which is trivial by the semisimplicity of L and hence ad is an isomorphism, as required. //

As shown in §6.1, a Lie group is compact if and only if it is a closed subgroup of some $O(m)$. Recalling that the Lie algebra of $O(m)$ is $\underline{\underline{o}}(m)$, it is not surprising, but nonetheless welcome, that a Lie algebra L is a Lie algebra of a compact Lie group if and only if it is a subalgebra of some $\underline{\underline{o}}(m)$. (Recall that a Lie algebra is said to be compact if it is the Lie algebra of some compact group.) This result is included in the following:

6.3.11 **Theorem.** (a) The following conditions on a (real, finite-dimensional) Lie algebra L are equivalent:

(i) L is compact;

(ii) L is a subalgebra of some $\underline{\underline{o}}(m)$; and

(iii) there exists an inner product $(X, Y) \mapsto F(X, Y)$ on L which satisfies

$$(6.3.2) \quad F([X, Y], Z) = F([Y, Z], X)$$

for all $X, Y, Z \in L$. (In the case of $\underline{\underline{o}}(m)$, this inner product can be taken as $(X, Y) \mapsto \mathrm{tr}(XY^T)$.)

(b) If L is a semisimple compact Lie algebra, then $(X, Y) \mapsto -K(X, Y)$ is an inner product on L.

6.3.12 **Remark.** If L is a complex simple Lie subalgebra of $\underline{\underline{gl}}(m, C)$, then Exercise 6.E shows that there exists $\lambda \in C$ such that

$$K(X, Y) = \lambda.\,\mathrm{tr}(XY) \quad \text{for } X, Y \in L.$$

Proof (of 6.3.11). (a) The proof that (i) implies (ii) is a combination of 5.1.2, 6.1.1 and the fact that the Lie algebra of $O(m)$ is $\underline{\underline{o}}(m)$, while the proof that (ii) implies (iii) is realised by noting that

$$F : (X, Y) \mapsto \mathrm{tr}(XY^T) = -\mathrm{tr}(XY)$$

is an inner product on each Lie subalgebra L of $\underline{\underline{o}}(m)$ and satisfies (6.3.2). (It is easily seen that F is symmetric, bilinear and satisfies (6.3.2). It is also positive definite for if $X = (x_{ij}) \in L$, then $\mathrm{tr}(XX^T) = \Sigma_{ij}(x_{ij})^2$ is positive and equals zero if and only if $X = 0$.)

To complete the cycle of implications, suppose that L is a Lie algebra with an inner product satisfying (6.3.2). Let L_0 denote the centre of L and let $L^\#$ denote the orthogonal complement of L_0 with respect to F. Then L_0 and $L^\#$ are ideals of L, $L_0 \cap L^\# = \{0\}$ and $L_0 \oplus L^\# = L$. Furthermore, $L^\#$ is semisimple. (Suppose that T is a commutative ideal in $L^\#$ and denote its orthogonal complement in $L^\#$ by $T^\perp$. Since

$$[T, L] = [T, L_o] + [T, T] + [T, T^{\perp}] = \{0\},$$

T is contained in the centre L_o of L which is possible only if $T = \{0\}$.)

The ideal L_o, being commutative, is clearly compact. (L_o is Lie isomorphic to $\mathbf{R}^k$, for some k, equipped with the trivial Lie product and hence is the Lie algebra of T^k. Cf. Exercises 2. A (ii), 3. A and 3. B.) In view of the proposition in 5. 2. 10 it remains only to show that $L^{\#}$ is compact. We will verify that this is indeed the case by showing that each semisimple Lie algebra L with an inner product F satisfying (6. 3. 2) is compact.

Define G to be the identity component of $\mathrm{Aut}(L)$; from Corollary 6. 3. 10 we know that the Lie algebra of G is L and hence it suffices to show that G is compact. A basic result established in Chapter 2 is that every connected Lie group H is generated by $\exp(\underline{\underline{h}})$. Thus, using 6. 3. 8 and the discussion in 6. 3. 9, G is generated by $\{e^{\operatorname{ad} X} : x \in L\}$. The bilinearity of F, the definition of $e^{(\cdot)}$ and the hypothesis (6. 3. 2) yield

$$\begin{aligned} F(e^{\operatorname{ad} X}Y, e^{\operatorname{ad} X}Z) &= \sum_{m,n=0}^{\infty} \frac{1}{m!\,n!} F((\operatorname{ad} X)^m Y, (\operatorname{ad} X)^n Z) \\ &= \sum_{m,n=0}^{\infty} \frac{(-1)^n}{m!\,n!} F((\operatorname{ad} X)^{m+n} Y, Z). \end{aligned}$$

For $k > 0$ the coefficient of $F((\operatorname{ad} X)^k Y, Z)$ is

$$\sum_{m+n=k} \frac{(-1)^n}{m!\,n!} = (k!)^{-1}(1 + (-1))^k = 0$$

and so $F(e^{\operatorname{ad} X}Y, e^{\operatorname{ad} X}Z) = F(Y, Z)$. Thus

$$F(D(Y), D(Z)) = F(Y, Z)$$

for all $D \in G$ and $Y, Z \in L$ showing that G is contained in $O(F)$, the group of all linear transformations of L which are orthogonal with respect to F. But G is closed and $O(F)$ is compact, in both cases with respect to the topology induced from $GL(L)$, and hence G is compact, as required.

(b) Let L be a compact Lie algebra and let F be an inner product on L satisfying (6.3.2), the existence of which is assured by part (a). Let $X_1, \ldots, X_n$ be an orthonormal basis of L with respect to F. For $X \in L$ write

$$[X, X_i] = \sum_{j=1}^{n} r_{ij}(X) X_j,$$

where the $r_{ij}(X)$ are real numbers. Since

$$F([X, X_i], X_j) = -F([X, X_j], X_i)$$

we have $r_{ij}(X) = -r_{ji}(X)$. Now

$$(\mathrm{ad}\, X)^2 X_i = \Sigma_{j,k}\, r_{ij}(X) r_{jk}(X) X_k$$

and so

$$\mathrm{tr}((\mathrm{ad}\, X)^2) = \Sigma_{i,j}\, r_{ij}(X) r_{ji}(X) = -\Sigma_{i,j} r_{ij}(X)^2.$$

Hence $\mathrm{tr}((\mathrm{ad}\, X)^2)$ is strictly negative whenever not all the $r_{ij}(X)$ are zero, that is, whenever ad X is nonzero. However, it was shown in the proof of 6.3.10 that $X \mapsto \mathrm{ad}\, X$ is injective whenever L is semisimple. Thus

$$-K(X, X) = -\mathrm{tr}((\mathrm{ad}\, X)^2) > 0 \quad \text{for} \quad X \neq 0$$

whenever L is compact semisimple, and hence -K is an inner product on L. //

As an immediate corollary we have that every Lie subalgebra of a compact Lie algebra must itself be compact.

Exercise 5.B demonstrated that a Lie subgroup of a compact group need neither be compact nor have the topology induced from the containing group. However, this is definitely not the case for semisimple Lie subgroups.

6.3.13 **Theorem.** Let G be a compact Lie group and $j : H \to G$ a semisimple connected Lie subgroup of G. Then (i) H, considered as a subset of G, is closed in G and (ii) the topology of H, considered as a Lie subgroup of G, is the induced topology.

Proof. If $\underline{\underline{h}}$ denotes the Lie algebra of H, then Theorem 6.3.11 shows that $\underline{\underline{h}}$ is compact since it is a subalgebra of compact $\underline{\underline{g}}$. Let K denote a compact semisimple Lie group which has $\underline{\underline{h}}$ as its Lie algebra. Let $\tilde{K}$ denote the simply-connected covering group of K; from 6.4.4 *infra*, $\tilde{K}$ is also compact. Since $\tilde{K}$ and H have the same Lie algebra, they are locally isomorphic. As explained in the last paragraph of 3.4.7, there exists an analytic homomorphism from $\tilde{K}$ onto H and so H is compact too.

We know that $j : H \to G$ is continuous and thus $j(H)$ must be compact, hence closed, in G. A similar argument shows that j maps each closed (and thus compact) subset of H onto a closed subset of G and hence the Lie subgroup topology of H is just that induced from G. //

This result shows that there is now no need to talk of semisimple connected Lie subgroups of a compact Lie group because all such subgroups are merely (semisimple connected) closed subgroups. Conversely, it was established in Chapter 3 that closed subgroups of a Lie group are Lie subgroups. This characterisation of semisimple connected Lie subgroups of a compact Lie group is basic to the proof of the first structure theorem in the next section.

6.4 The structure of compact Lie groups

The structure theorems presented in this and the following section can be said to be the culmination of these notes for in proving them we will need to bring together, both directly and indirectly, very nearly all of the preceding material. First a lemma.

6.4.1 **Lemma.** *If K is a totally disconnected normal subgroup of a connected topological group G, then K is contained in the centre of G.*

Proof. For each $x \in K$, define $B_x : G \to K$ by $y \mapsto yxy^{-1}$. Since each B_x is continuous, then each $B_x(G)$ is connected. But K is totally disconnected and so $B_x(G)$ must consist of precisely one point, namely

x. Hence $yxy^{-1} = x$ for all $x \in K$ and $y \in G$, showing that K is central, as asserted. //

We now describe the structure of a compact connected Lie group in terms of the identity component of its centre and of all its simple normal connected closed subgroups.

6.4.2 **Theorem.** Every compact connected Lie group G is topologically isomorphic with

$$(T_0 \times G_1 \times \ldots \times G_m)/K,$$

where T_0 is the identity component of the centre of G, the G_i are all the simple normal connected closed subgroups of G, and K is a finite subgroup of the centre of the product. Furthermore, $T_0 \cap K = \{e\} = G_i \cap K$ for $i \in \{1, \ldots, m\}$ or, more formally, $(T_0 \times \{e_1\} \times \ldots \times \{e_m\}) \cap K = \{e_0\} \times \ldots \times \{e_m\}$ and similarly for the G_i.

Remark. If H is a compact Lie group and K a closed normal subgroup then G/K is also a compact Lie group by 6.1.3 (b). Invoking this result and combining it with Corollary 2.3.2 shows that G and $(T_0 \times G_1 \times \ldots \times G_m)/K$ are isomorphic as Lie groups. However, in our case K is finite and hence the fact that $(T_0 \times G_1 \times \ldots \times G_m)/K$ is a Lie group can be arrived at more simply by noting that it is locally isomorphic with the Lie group $T_0 \times G_1 \times \ldots \times G_m$.

Proof (of 6.4.2). Let G be a compact connected Lie group and let $\mathrm{Ad} : G \to GL(\underline{\underline{g}})$ denote the adjoint representation of G. Since G is compact we can choose an inner product on $\underline{\underline{g}}$ so that Ad is an orthogonal representation. That is, $\mathrm{Ad}(G) \subseteq O(\underline{\underline{g}})$ (see A.3).

Section A.2.3 affirms that $\underline{\underline{g}}$ may be written as a direct sum of vector spaces

$$(6.4.1) \quad \underline{\underline{g}} = V_0 \oplus V_1 \oplus \ldots \oplus V_m,$$

where

$$V_0 = \{X \in \underline{\underline{g}} : \mathrm{Ad}(x)X = X \text{ for all } x \in G\}$$

and the V_j $(j = 1, 2, \ldots, m)$ are minimal subspaces of $\underline{\underline{g}}$ invariant under Ad. Define $\gamma_j = \mathrm{Ad}|V_j$ for $j = 1, 2, \ldots, m$. Then each γ_j is a nontrivial irreducible continuous representation of G. Since V_j $(j = 0, 1, \ldots, m)$ is invariant under the action of Ad, they must be ideals of $\underline{\underline{g}}$ by Lemma 5.2.5. Let G_j [resp. T_0] denote the unique connected normal Lie subgroup of G corresponding to V_j for $j = 1, \ldots, m$ [resp. V_0] as in Theorem 5.2.6.

Fact 1. The ideal V_0 is the centre of $\underline{\underline{g}}$ and T_0 is the connected component of the identity of the centre of G.

This is Corollary 5.2.8.

Fact 2. The G_j $(j = 1, \ldots, m)$ consist of all the simple connected normal Lie subgroups of G and hence, by Theorems 3.3.1 and 6.3.13, of all the simple connected normal closed subgroups of G. (See the remarks following the proof of Theorem 6.3.13.)

Each representation γ_j $(j = 1, 2, \ldots, m)$ is irreducible, whence the corresponding V_j, and hence the corresponding G_j, are simple. Now suppose that $j : H \to G$ is a simple connected normal Lie subgroup of G, and let $\underline{\underline{h}}$ denote the Lie algebra of H. By Theorem 5.2.6, $\underline{\underline{h}}$ is a (simple) ideal in $\underline{\underline{g}}$. The decomposition (6.4.1) satisfies the requirements of Lemma 6.3.2 (iii), and so $\underline{\underline{h}} = V_k$ for some $k \in \{1, 2, \ldots, m\}$, whence $H = G_k$.

Fact 3. Whenever $x \in G_j$, $y \in G_k$, where $j, k \in \{1, 2, \ldots, m\}$ and $j \neq k$, then $xy = yx$.

Let V be the image under the exponential map of a neighbourhood of 0 in $\underline{\underline{g}}$ on which exp is injective. Since $V_j \cap V_k = \{0\}$, then $G_j \cap G_k \cap V = \{e\}$. For $x \in G_j$ and $y \in G_k$, the normality of G_j and G_k implies that $xyx^{-1}y^{-1} \in G_j \cap G_k$. Suppose also that x, y belong to some neighbourhood W of e in G which ensures that $xyx^{-1}y^{-1} \in V$. Then $xyx^{-1}y^{-1} = \{e\}$, that is $xy = yx$. Now G_j and G_k are connected and so are generated by $G_j \cap W$ and $G_k \cap W$ respectively. (Use 2.1.5 and the fact that the topology of a Lie subgroup is stronger than the induced topology.) Thus $xy = yx$ for all $x \in G_j$, $y \in G_k$.

<u>Fact 4.</u> The map $\phi : (x_0, x_1, \ldots, x_m) \mapsto x_0 x_1 \ldots x_m$ is a local diffeomorphism and an analytic homomorphism from $T_0 \times G_1 \times \ldots \times G_m$ onto G. Also its kernel K is a finite subset of the centre of $T_0 \times G_1 \times \ldots \times G_m$.

In view of facts 1 and 3, the map ϕ is a homomorphism. It is also continuous and hence analytic. The other required properties of ϕ will be made apparent by describing it explicitly within a certain neighbourhood of the identity of $T_0 \times G_1 \times \ldots \times G_m$. We first consider $\underline{\underline{g}} = V_0 \oplus V_1 \oplus \ldots \oplus V_m$ as the Lie algebra of G. Then since $[V_i, V_j] = \{0\}$ from pairs $i \neq j$, the CBH formula for Lie groups allows us to write

$$\exp(X_0 + X_1 + \ldots + X_m) = \exp X_0 \exp X_1 \ldots \exp X_m,$$

whenever $X_i \in V_i$, $i = 0, 1, \ldots, m$. Alternatively, if $\underline{\underline{g}} = V_0 \oplus V_1 \oplus \ldots \oplus V_m$ is taken as the Lie algebra of $T_0 \times G_1 \times \ldots \times G_m$, Proposition 5.2.10 shows that

$$\exp_{T_0 \times G_1 \times \ldots \times G_m}(X_0 + X_1 + \ldots + X_m) = (\exp_{T_0} X_0, \exp_{G_1} X_1, \ldots, \exp_{G_m} X_m).$$

Thus, by working within suitable neighbourhoods of the identities of $T_0 \times G_1 \times \ldots \times G_m$ and G, neighbourhoods on which $\log_{T_0 \times G_1 \times \ldots \times G_m}$ and $\log_G$ are defined as the local inverses of $\exp_{T_0 \times G_1 \times \ldots \times G_m}$ and $\exp_G$, we have

$$\begin{aligned} x_0 x_1 \ldots x_m &= (\exp_G \log_G x_0) \ldots (\exp_G \log_G x_m) \\ &= \exp_G(\log_G x_0 + \ldots + \log_G x_m) \end{aligned}$$

$$= \exp_G \circ \log_{T_0 \times G_1 \times \ldots \times G_m}(\exp_{T_0} \log_G x_0, \exp_{G_1} \log_G x_1, \ldots, \exp_{G_m} \log_G x_m)$$

$$= \exp_G \circ \log_{T_0 \times G_1 \times \ldots \times G_m}(x_0, x_1, \ldots, x_m),$$

the last line following from the fact that $\exp_{T_0}$, $\exp_{G_1}$, etc., is precisely the restriction of $\exp_G$ to V_0, V_1, etc. (See the remarks at the end of §5.1.) Thus there exists a neighbourhood of the identity of $T_0 \times G_1 \times \ldots \times G_m$ on which ϕ is a diffeomorphism since it is equal to $\exp_G \circ \log_{T_0 \times G_1 \times \ldots \times G_m}$ locally.

In this case the kernel K of ϕ is discrete, hence finite (since $T_0 \times G_1 \times \ldots \times G_m$ is compact) and totally disconnected. It is also normal and so, by Lemma 6.4.1, it must be central. That ϕ is surjective follows the facts that it is a homomorphism and a local diffeomorphism, and that G is connected (use 2.1.5).

We now conclude the proof. Since $\phi(x_0, e_1, \ldots, e_m) = x_0$, it follows that $T_0 \cap K = \{e\}$. Similarly, $G_i \cap K = \{e\}$ for $i = 1, 2, \ldots, m$. Since ϕ is a local diffeomorphism it is certainly open (and continuous). Also, as remarked above, it is surjective and so an application of the second isomorphism theorem for topological groups (Theorem (5.34) of Hewitt and Ross [1]) shows that G and $(T_0 \times G_1 \times \ldots \times G_m)/K$ are topologically isomorphic. //

6.4.3 **Compact semisimple Lie groups.** The results promised in 6.3.3 are attainable from Facts 1 and 2 of the above proof. For if $\underline{\underline{g}}$ is a compact Lie algebra, then we have seen that it has a decomposition $\underline{\underline{g}} = V_0 \oplus V_1 \oplus \ldots \oplus V_m$, where V_0 is the centre of $\underline{\underline{g}}$ and the V_i $(i = 1, \ldots, m)$ are all the simple ideals of $\underline{\underline{g}}$. Thus a compact Lie algebra is semisimple if and only if it has a trivial centre. Consequently a compact connected Lie group is semisimple if and only if the identity component of its centre is trivial, this being the case if and only if it has a finite centre.

The following is a rather difficult result which will enable the above structure theorem to be stated with simply-connected subgroups in place of connected ones. (For a proof of this result, see Hausner and Schwartz [1, p. 221], Helgason [1, p. 123] or Hochschild [1, p. 144].)

6.4.4 **Theorem.** <u>The covering group of a compact connected semisimple Lie group is compact.</u>

The hypothesis of semisimplicity is essential here, for if G is a compact connected Lie group which is not semisimple, then from the above structure theorem G is locally isomorphic with $T^k \times H$, where $k > 0$ and H is a compact connected semisimple group. The covering groups of G and $T^k \times H$ are the same, and since the covering group of T^k is $\mathbf{R}^k$, the covering group of G cannot be compact. (The covering group of a product is the product of the covering groups.)

6.4.5 **Theorem.** Every compact connected Lie group G is topologically isomorphic (even analytically isomorphic) with a unique group of the form

$$(T_0 \times \tilde{G}_1 \times \dots \times \tilde{G}_m)/K,$$

where T_0 is the identity component of the centre of G, the $\tilde{G}_i$ are compact simple simply-connected Lie groups, and K is a finite subgroup of the centre of the product.

Proof. First let us deal with the uniqueness of such a description. Whenever G has the form described in the statement of the theorem, we know from the proof of Theorem 6.4.2 that we have a decomposition of its Lie algebra of the form $V_0 \oplus V_1 \oplus \dots \oplus V_m$ where V_0 is the centre of $\underline{g}$ and the V_j, $j = 1, \dots, m$, are simple ideals of $\underline{g}$. But such a decomposition has been shown to be unique and hence the uniqueness of $\tilde{G}_1, \dots, \tilde{G}_m$ (and of course, T_0) follows because a simply-connected Lie group is determined entirely by its Lie algebra.

Turning to existence, let T_0 denote the identity component of the centre of G, and G_i, $i = 1, 2, \dots, m$, the simple normal connected closed subgroups of G. The map of $\phi : T_0 \times G_1 \times \dots \times G_m \to G$ defined by $\phi(x_0, x_1, \dots, x_m) = x_0 x_1 \dots x_m$ was shown in the proof of 6.4.2 to be an analytic epimorphism and a local isomorphism. Using the notation of 3.4.7, let $p_i : \tilde{G}_i \to G_i$, $i = 1, \dots, m$, denote the simply-connected covering group of G_i. Each p_i is an analytic epimorphism and a local isomorphism, and each $\tilde{G}_i$ is compact by 6.4.4. Define

$$p : T_0 \times \tilde{G}_1 \times \dots \times \tilde{G}_m \to G \text{ by } p(x_0, x_1, \dots, x_m) = \phi(x_0, p_1 x_1, \dots, p_m x_m).$$

Then p is an analytic epimorphism and a local isomorphism, and the proof is completed analogously to the completion of the proof of 6.4.2. //

6.4.6 **Simple compact Lie groups.** The usefulness of the above structure theorems is enhanced considerably when it is realised that all the simple simply-connected compact groups, and their centres, can be fully described. This description is carried out via an analysis of all simple Lie algebras ... here we will just quote the pertinent results (but for details see Bourbaki [1, Chap. VI], Jacobson [1] or Samelson [1]). We first describe some classes of groups.

(i) Let A_k denote the local isomorphism class of $SU(k+1)$, $k = 1, 2, \ldots$.

(ii) Let B_k denote the local isomorphism class of $SO(2k+1)$ and D_k the local isomorphism class of $SO(2k)$, $k = 1, 2, \ldots$.

(iii) Let C_k denote the local isomorphism class of the symplectic group $Sp(k)$, $k = 1, 2, \ldots$. (The symplectic group is defined as follows: let Q^n denote quaternionic n-space equipped with the 'inner product'

$$(x, y) = \sum_{m=1}^{n} x_m \bar{y}_m,$$

where $x = (x_1, \ldots, x_n)$, $y = (y_1, \ldots, y_n) \in Q^n$ and if $u = a + bi + cj + dk \in Q$, then $\bar{u} = a - bi - cj - dk$. Then $Sp(n)$ is defined as the group of transformation of Q^n, linear over Q, which are invariant under the above inner product.)

The above classes are said to be 'classical' in contradistinction with:

(iv) the five so-called 'exceptional' complex Lie algebras; the local isomorphism classes of Lie groups corresponding to them are commonly denoted by E_6, E_7, E_8, F_4 and G_2. (Their descriptions will not be given here, however.)

In each case the subscript gives the rank of the (class of) groups which is defined as the dimension of the largest torus that can be embedded in it. For example, the rank of A_k is k while the rank of E_6 is 6. Also the collection A_k $(k \geq 1)$, B_k $(k \geq 2)$, C_k $(k \geq 3)$, D_k $(k \geq 4)$, E_6, E_7, E_8, F_4 and G_2 is precisely the set, without repetition, of all the local isomorphism classes of the compact simple Lie groups. Some redundancies occur in the lower values of the rank:

$A_1 = B_1 = C_1$
$D_2 = A_1 \times A_1$
$B_2 = C_2$
$A_3 = D_3$
D_1 has **T** as a representative.

The groups SU(n) and Sp(n) are already simply-connected. The fundamental group of SO(n) for $n \geq 3$ is $\mathbf{Z}_2$. (Whenever n is a positive integer, $\mathbf{Z}_n$ will denote the group of integers modulo n.) The simply-connected covering group of SO(n) is called the spinor group and is denoted by Spin(n). The following list gives the dimensions and describes the centres of the simply-connected representatives $\tilde{G}$ of the local isomorphism classes of the simple compact groups given in (i)-(iv) above.

Class	dimension	centre of $\tilde{G}$
A_k	$k(k+2)$	$\mathbf{Z}_{k+1}$
B_k	$k(2k+1)$	$\mathbf{Z}_2$
C_k	$k(2k+1)$	$\mathbf{Z}_2$
D_{2k}	$2k(4k-1)$	$\mathbf{Z}_2 \oplus \mathbf{Z}_2$
D_{2k+1}	$(2k+1)(4k+1)$	$\mathbf{Z}_4$
E_6	78	$\mathbf{Z}_3$
E_7	133	$\mathbf{Z}_2$
E_8	248	$\{0\}$
F_4	52	$\{0\}$
G_2	14	$\{0\}$

6.5 Compact connected groups

In this section we first establish the well-known result that every compact group is a projective limit of compact Lie groups. This will then be used to prove the lesser-known result due to Pontrjagin [2] and van Kampen [1] that every compact connected group is topologically isomorphic to $(T \times \Pi_{i \in I} G_i)/K$, where T is the identity component of the centre of G, $(G_i : i \in I)$ is a family of simply connected, simple, compact Lie groups and K is a closed, totally disconnected subgroup of the centre of the total product. This result reduces to Theorem 6.4.5 when G is also Lie.

6.5.1 **Projective limits.** Let A be a directed partially ordered set in which the order relation is denoted by '$\leq$'. For each $\alpha \in A$ let G_α be a topological group and for each pair (α, β) with $\alpha < \beta$ (that is, with $\alpha \leq \beta$ and $\alpha \neq \beta$) suppose that there exists a continuous homomorphism $f_{\beta\alpha}$ from G_β into G_α. Suppose also that

(6.5.1) $f_{\gamma\alpha} = f_{\beta\alpha} \circ f_{\gamma\beta}$ whenever $\alpha < \beta < \gamma$.

The collection consisting of A, the groups G_α, and the maps $f_{\beta\alpha}$ is called an <u>inverse mapping system.</u> Let G denote the subset of $\Pi_{\alpha \in A} G_\alpha$ consisting of those points x such that

(6.5.2) $pr_\alpha(x) = f_{\beta\alpha} pr_\beta(x)$ whenever $\alpha < \beta$,

where pr_γ denotes the natural projection onto G_γ. It is easily checked that G is closed under the group operations on $\Pi_\alpha G_\alpha$; when G is equipped with this group structure and with the topology induced from $\Pi_\alpha G_\alpha$, then it is called the <u>projective limit</u> (or <u>inverse limit</u>) of the above inverse system. (A categorical definition of these limits is given in Exercise 6.G.) Often we will take A and the mappings $f_{\beta\alpha}$ to be implicitly given and write $G = \varprojlim G_\alpha$.

The fact that the groups G_α fall under the blanket assumption stated in Chapter 2 that we only consider topological groups which are Hausdorff has the consequence that G is closed in $\Pi_\alpha G_\alpha$. Also a simpler description of the topology of G can be derived from the directedness of A, namely that sets of the form

(6.5.3) $(pr)_\alpha^{-1} U_\alpha$,

where $\alpha \in A$ and U_α is an open neighbourhood of the identity in G_α, form an open basis of neighbourhoods of the identity in G. (Exercise.)

6.5.2 **Example.** Let $(G_i : i \in I)$ be a family of topological groups and let A be the set of finite subsets of I partially ordered by $\alpha \geq \beta$ if and only if $\alpha \supseteq \beta$. Define $G^\alpha = \Pi_{i\in\alpha} G_i$ and $f_{\beta\alpha} : G^\beta \to G^\alpha$ for $\beta > \alpha$ as the natural projection. Then A, the G^α and the $f_{\beta\alpha}$ form an inverse mapping system; let $G = \varprojlim G^\alpha$.

Define a map $\phi : G \to \Pi_{i\in I} G_i$ in the following manner: Whenever $x \in G$ and $i \in I$, then $\{i\} \in A$. Define $\phi(x) = (\mathrm{pr}_{\{i\}} x)_{i\in I}$. Then ϕ is a homomorphism. Also it is injective since $\mathrm{pr}_{\{i\}} = f_{\alpha,\{i\}} \mathrm{pr}_\alpha$ whenever $i \in \alpha$. Finally notice that it is surjective and a homeomorphism and hence that G and $\Pi_{i\in I} G_i$ are topologically isomorphic. Thus a <u>direct product of topological groups is a special case of a projective limit.</u>

6.5.3 **Compact groups as projective limits.** (i) Let G be a compact group and let A be a maximal collection of pairwise inequivalent continuous unitary representations of G. For $\alpha \in A$, put $K_\alpha = \ker(\alpha)$ and define a partial order on A by $\alpha \leq \beta$ if and only if $K_\alpha \supseteq K_\beta$. Given $\alpha, \beta \in A$ there exists $\gamma \in A$ such that γ and $\alpha \oplus \beta$ are equivalent. Hence A is directed since $K_\gamma = K_{\alpha\oplus\beta} \subseteq K_\alpha \cap K_\beta$. For $\alpha \in A$ define $G_\alpha = G/K_\alpha \cong \alpha(G)$; each G_α is topologically isomorphic with a closed subgroup of $U(d(\alpha))$ and so must be a compact Lie group (Theorem 6.1.1). Also notice that the G_α are connected whenever the same is true of G. Whenever $\alpha < \beta$, define $f_{\beta\alpha} : G_\beta \to G_\alpha$ by $f_{\beta\alpha}(xK_\beta) = xK_\alpha$ for all $xK_\beta \in G_\beta$. Then A, the groups G_α and the mappings $f_{\beta\alpha}$ form an inverse mapping system (notice that not only are the $f_{\beta\alpha}$ continuous, but they are also open and surjective). Define a homomorphism

$$\phi : G \to \Pi_\alpha G_\alpha \text{ by } \phi(x) = (xK_\alpha)_{\alpha\in A}.$$

Some routine manipulations show that the range of ϕ is contained in $\varprojlim G_\alpha$, that it is continuous and that it is injective (use A.2.2). To see that ϕ actually has $\varprojlim G_\alpha$ as its range we must show that corresponding to any $(x_\alpha K_\alpha)_{\alpha\in A} \in \varprojlim G_\alpha$ there exists $x \in G$ such that $x \in \cap_\alpha x_\alpha K_\alpha$, for then $\phi(x) = (x_\alpha K_\alpha)_{\alpha\in A}$. Given $\alpha_1, \ldots, \alpha_n \in A$ there exists

$\beta \in A$ with $\beta \geq \alpha_i$ for all i. Hence $x_\beta K_\beta \subseteq x_\beta K_{\alpha_i} = f_{\beta\alpha_i}(x_\beta K_\beta) = x_{\alpha_i} K_{\alpha_i}$. Thus $x_\beta K_\beta \subseteq \cap_{i=1}^n x_{\alpha_i} K_{\alpha_i}$ so that the family $\{x_\alpha K_\alpha\}_{\alpha \in A}$ of closed sets has the finite intersection property. In this case the full intersection of these sets must be nonempty because G is compact. Finally use the open mapping theorem (A. 1. 5) or merely the compactness of G. Thus G is a projective limit of compact Lie groups.

Two additional facts which will be useful later are that the compact Lie groups in the projective limit can be taken as connected whenever G is also connected and that the homomorphisms in the inverse mapping system can be taken as continuous open surjections. Also, whenever U is an open neighbourhood of e in G, then we can find a closed normal subgroup K of G such that G/K is a compact Lie group. This follows from the simplification of the topology mentioned at the end of 6. 5. 1 because this provides the assurance that there exists $\beta \in A$ and an open neighbourhood U_β of e_β in $\beta(G)$ such that $\phi(U) \supseteq (pr_\beta)^{-1} U_\beta$. But this last set contains

$$(pr_\beta)^{-1} e_\beta = \{(xK_\alpha)_{\alpha \in A} \in \varprojlim G_\alpha : x \in K_\beta\} = \phi(K_\beta).$$

Thus $U_\beta \supseteq K_\beta$ and clearly K_β satisfies the sought-after conditions.

In the remainder of 6. 5. 3 we suppose that G is a compact group and that A, G_α, $f_{\beta\alpha}$ and ϕ are as in (i). Denote by f_α the canonical projection $f_\alpha : G \rightarrow G/K_\alpha = G_\alpha$.

(ii) Let H be a closed subgroup of G and for each $\alpha \in A$ define $H_\alpha = f_\alpha(H)$. Then H_α is the continuous image of a compact set and so is itself compact, hence closed, in G_α. It is clear that whenever $\alpha < \beta$, then the restriction of $f_{\beta\alpha}$ to H_β is a continuous epimorphism from H_β onto H_α. Hence we have another inverse mapping system. We will see now that the restriction of the isomorphism $\phi : G \rightarrow \varprojlim G_\alpha$ to H is a topological isomorphism from H onto $\varprojlim H_\alpha$.

From the definition of ϕ, ϕ is a topological isomorphism from the subgroup

$$\cap_{\alpha \in A} HK_\alpha \text{ of } G \text{ onto } \varprojlim H_\alpha.$$

The proof is complete provided we can show that $\cap_\alpha HK_\alpha = H$. This follows from the general result that if V is any closed subset of G, then

$$(6.5.4) \quad \cap_{\alpha\in A} VK_\alpha = \cap_{\alpha\in A} K_\alpha V = V.$$

The normality of K_α ensures the equality of the first two sets while it is evident that they both contain V. On the other hand, since V is closed, $x \in V$ if and only if xU intersects V for every open symmetric neighbourhood U of e. This is the case if and only if $x \in VU$ for all such neighbourhoods. The final condition is satisfied by $x \in \cap_\alpha VK_\alpha$.

(iii) Suppose that H is a closed subgroup of G. Then $H = \varprojlim H_\alpha$ is connected if (and only if) each $H_\alpha = f_\alpha(H)$ is also connected. (This follows from (ii) once we show that $f_\alpha(C) = H_\alpha$ for each α since C is also a closed subgroup of G. The compactness of C implies that $f_\alpha(C)$ is closed in H_α. Furthermore, the restriction of f_α to H is a continuous open surjection onto H_α and so by Theorem 7.12 of Hewitt and Ross [1], $f_\alpha(C)$ is the identity component in H_α, that is, $f_\alpha(C) = H_\alpha$, as required. (Theorem 7.12 of Hewitt and Ross [1] states: 'Let G be a locally compact group and let C be the component of the identity in G. Let f be an open, continuous homomorphism of G onto a topological group $G^\#$, and let $C^\#$ be the component of the identity in $G^\#$. Then $f(C)^- = C^\#$.'))

(iv) (Converse to (ii).) For each $\alpha \in A$ suppose that H_α is a closed subgroup of G such that $f_{\beta\alpha}(H_\beta) = H_\alpha$ whenever $\alpha < \beta$. Then there is a unique closed subgroup H of G such that $f_\alpha(H) = H_\alpha$.

Proof. Let $g_{\beta\alpha}$ be the restriction of $f_{\beta\alpha}$ to H_β. The set A, the groups H_β and the maps $g_{\beta\alpha}$ form an inverse mapping system and it follows immediately that $\varprojlim H_\alpha$ is equal to $\phi(G) \cap \Pi_\alpha H_\alpha$. Denote the inverse image of $\varprojlim H_\alpha$ in G via ϕ by H. Thus

$$(6.5.5) \quad H = \cap_{\alpha\in A} f_\alpha^{-1}(H_\alpha);$$

we will now sketch the proof that $f_\alpha(H) = H_\alpha$ for each α.

Let Σ be the set of families $\mathfrak{U} = (U_\alpha)_{\alpha \in A}$ which satisfy:

$U_\alpha \neq \emptyset$ and U_α is closed in G_α for all $\alpha \in A$;

$f_{\beta\alpha}(U_\beta) \subseteq U_\alpha$ whenever $\alpha \leq \beta$.

(The symbol $f_{\alpha\alpha}$ will always denote the identity map on G_α.) Partially order Σ by the relation $\mathfrak{U} \leq \mathfrak{V}$ whenever $U_\alpha \supseteq V_\alpha$ for all α, where $\mathfrak{U} = (U_\alpha)$ and $\mathfrak{V} = (V_\alpha)$.

If $\mathfrak{U}^\lambda = (U_\alpha^\lambda)$ is a chain in Σ, then $\mathfrak{U} = (\cap_\lambda U_\alpha^\lambda)_{\alpha \in A}$ is in Σ (use the compactness of G) and is an upper bound to the chain. Thus Zorn's lemma applies and Σ has at least one maximal element. Whenever $\mathfrak{U} = (U_\alpha)$ is a maximal element, then

$$U_\alpha = f_{\beta\alpha}(U_\beta) \text{ for all } \alpha \leq \beta.$$

(For each α, define $V_\alpha = \cap_{\beta \geq \alpha} f_{\beta\alpha}(U_\beta)$. Then $\mathfrak{V} = (V_\alpha) \in \Sigma$ and $\mathfrak{V} \geq \mathfrak{U}$. Now use the maximality of $\mathfrak{U}$.) Furthermore, each U_α must be a singleton. (Select $\alpha' \in A$ and then $x_{\alpha'} \in U_{\alpha'}$. For $\beta \geq \alpha'$ put $V_\beta = U_\beta \cap f_{\beta\alpha'}^{-1}(x_{\alpha'})$, otherwise put $V_\beta = U_\beta$. Now $\mathfrak{V} = (V_\alpha)_{\alpha \in A} \in \Sigma$. Clearly $\mathfrak{V} \geq \mathfrak{U}$ and so they are equal because $\mathfrak{U}$ is maximal. Thus $V_\beta = U_\beta$ for all β and in particular, $U_{\alpha'} = \{x_{\alpha'}\}$.)

To complete the proof that $f_\alpha : H \to H_\alpha$ is surjective it is necessary to show that $f_\alpha(H) \supseteq H_\alpha$ for each α. Let $x_{\alpha'} \in H_{\alpha'}$ for fixed $\alpha' \in A$ and put $U_\beta = f_{\beta\alpha}^{-1}(x_{\alpha'})$ if $\beta \geq \alpha'$ and $U_\beta = H_\beta$ otherwise. Then $\mathfrak{U} = (U_\beta)_{\beta \in A} \in \Sigma$. Let $\mathfrak{V} = (V_\beta)_{\beta \in A}$ be a maximal element of Σ with $\mathfrak{V} \geq \mathfrak{U}$. Each V_β must be of the form $\{y_\beta\}$ and so $y = (y_\beta) \in \varprojlim H_\alpha$. Thus $\phi^{-1}(y) \in H$ and

$$f_{\alpha'}(\phi^{-1}(y)) = y_{\alpha'} = x_{\alpha'}.$$

If $H^\#$ is another closed subgroup of G with $f_\alpha(H^\#) = H_\alpha$ for all $\alpha \in A$, then using (6.5.4) and (6.5.5) we have:

$$H^\# = \cap_\alpha H^\# K_\alpha = \cap_\alpha f_\alpha^{-1}(H_\alpha) = H. \quad //$$

(v) The compact group G is connected and Abelian if and only if all the groups G_α are torii. (Use Exercise 3. B and part (iii) above.)

Before proving the promised structure theorem, we need the following lemma characterising epimorphisms between decomposable Lie algebras.

6.5.4 **Lemma.** Suppose that L and M are Lie algebras with decompositions

$$L_0 \oplus L_1 \oplus \ldots \oplus L_m \text{ and } M_0 \oplus M_1 \oplus \ldots \oplus M_n$$

respectively, where L_0 [resp. M_0] denotes the centre of L [resp. M] and the L_i, $1 \leq i \leq m$ [resp. the M_j, $1 \leq j \leq n$] are all the simple ideals of L [resp. M] (see 6.3.2 (iii)). Also suppose that ϕ is a Lie epimorphism from L onto M. Then

(i) $\phi(L_0) = M_0$;

(ii) for each $i \in \{1, \ldots, m\}$, either there exists an integer, $p(i)$ say, in $\{1, \ldots, n\}$ such that ϕ is a Lie isomorphism from L_i onto $M_{p(i)}$ or $\phi(L_i) = \{0\}$;

(iii) for each $j \in \{1, \ldots, n\}$ there exists a unique $i \in \{1, \ldots, m\}$ such that $p(i) = j$.

Proof. Let us begin with the proof of (ii). Given $i \in \{1, \ldots, m\}$, $\phi(L_i)$ is an ideal of M since

$$[\phi(L_i), M] = [\phi(L_i), \phi(L)] = \phi([L_i, L]) \subseteq \phi(L_i).$$

Suppose that K is an ideal of $\phi(L_i) \neq \{0\}$ and define $P = \{X \in L_i : \phi(X) \in K\}$. Then P is an ideal of L_i since $[P, L_i] \subseteq L_i$ and

$$\phi([P, L_i]) = [\phi(P), \phi(L_i)] = [K, \phi(L_i)] \subseteq K.$$

But L_i is simple so that $P = \{0\}$ or L_i in which case $K = \{0\}$ or $\phi(L_i)$ respectively. In other words, $\phi(L_i)$ is a simple ideal in M and so must be equal to some M_j. Denote this j by $p(i)$. A consequence of the simplicity of L_i is that ϕ restricted to L_i is an isomorphism from L_i onto $M_{p(i)}$.

Turning to (i), $\phi(L_0)$ is a subset of M_0, the centre of M, because

$$[\phi(L_0), M] = [\phi(L_0), \phi(L)] = \phi([L_0, L_0]) \\ = \phi(\{0\}) = \{0\}.$$

But since ϕ is surjective we must have that $\phi(L_0) = M_0$.

The fact that to each $j \in \{1, \ldots, n\}$ there exists at least one $i \in \{1, \ldots, m\}$ with $p(i) = j$ follows from the surjectivity of ϕ. Now suppose that (iii) is not valid. By renumbering if necessary, it follows that $p(1) = p(2) = 1$. Let

$$K = \{(X, Y) \in L_1 \oplus L_2 : \phi(X, Y) = 0\}.$$

It follows that $(X, Y) \in K$ if and only if $\phi(X) = -\phi(Y)$ and so, since ϕ is an isomorphism on both L_1 and L_2, the natural projection of K onto L_1 is also an isomorphism. Thus K is another simple ideal of $L_1 \oplus L_2$, an impossibility since L_1 and L_2 are the only simple ideals in $L_1 \oplus L_2$ (see 6.3.2 (iii)). This proves (iii) and so completes the proof. //

6.5.5 **Remark.** Suppose that L, M and ϕ are as in the statement of the lemma but that L and M both have trivial centres (which is the case if and only if they are semisimple). By identifying each simple ideal of M with its isomorph in L via ϕ, we see that ϕ is nothing but a natural projection from a direct sum of simple ideals onto a partial sum.

6.5.6 **Structure of compact groups.** Let T be a compact connected Abelian group, $(G_i : i \in I)$ a family of compact, simply connected, simple Lie groups and K a totally disconnected, closed subgroup of the centre of $T \times \Pi_i G_i$. Then

$$G = (T \times \Pi_{i \in I} G_i)/K$$

is a compact connected group and every compact connected group is of this form. Furthermore, given a compact connected group G, we can always choose T, G_i $(i \in I)$ and K as above with the properties that T is the identity component of the centre of G, and $K \cap T = \{e\}$.

Proof. With T, G_i and K as in the statement of the theorem it is trivial to verify that $(T \times \Pi_i G_i)/K$ is compact and connected. Now suppose that G is a compact connected group. From 6.5.3 we know that G is the limit of an inverse mapping system consisting of a directed set A, compact connected Lie groups G_α, and continuous open epimorphisms $f_{\beta\alpha}$. As in the proof of 6.4.2, decompose the Lie algebra $\underline{\underline{g}}_\alpha$ of each G_α in the form

$$(6.5.6) \quad \underline{\underline{g}}_\alpha = L_{\alpha 0} \oplus L_{\alpha 1} \oplus \ldots \oplus L_{\alpha m(\alpha)},$$

where $L_{\alpha 0}$ is the centre of $\underline{\underline{g}}_\alpha$ and the $L_{\alpha i}$ $(1 \le i \le m(\alpha))$ are all the simple ideals of $\underline{\underline{g}}_\alpha$. Let T_α denote the identity component of the centre of G_α (and so the Lie algebra of T_α is $L_{\alpha 0}$) and Σ_α the unique connected closed semisimple subgroup of G_α with $L_{\alpha 1} \oplus \ldots \oplus L_{\alpha m(\alpha)}$ as its Lie algebra.

By arguing as in Fact 4 of the proof of Theorem 6.4.2 it is clear that the map from $T_\alpha \times \Sigma_\alpha$ to G_α defined by $(t, \sigma) \mapsto t\sigma$ is a local diffeomorphism and an epimorphism with a finite Abelian kernel, P_α say. From A.3.9 we know that P_α is isomorphic with $T_\alpha \cap \Sigma_\alpha$ and that $P_\alpha \cap T_\alpha = \{e\} = P_\alpha \cap \Sigma_\alpha$ (more precisely, $P_\alpha \cap (T_\alpha \times \{e\}) = \{e\}$, etc.). Also G_α and $(T_\alpha \times \Sigma_\alpha)/P_\alpha$ are topologically isomorphic.

If $\alpha < \beta$ and $f_{\beta\alpha}$ is the corresponding inverse mapping, then $(f_{\beta\alpha})_{*,e}$ satisfies the hypotheses of Lemma 6.5.4 and so the appropriate restriction of $f_{\beta\alpha}$ are continuous open epimorphisms from T_β onto T_α and from Σ_β onto Σ_α. In this way $(T_\alpha)_{\alpha \in A}$ and $(\Sigma_\alpha)_{\alpha \in A}$ form inverse mapping systems. Making use of the results in 6.5.3, let ϕ denote the isomorphism

$$\phi : G \to \varprojlim G_\alpha \quad \text{where } \phi(x) = (xK_\alpha)_{\alpha \in A},$$

and let T and Σ denote the unique closed subgroups of G which are ϕ-isomorphic with $\varprojlim T_\alpha$ and $\varprojlim \Sigma_\alpha$ respectively. Then T and Σ are closed connected subgroups of G satisfying

$$f_\alpha(T) = T_\alpha, \quad f_\alpha(\Sigma) = \Sigma_\alpha$$

for each $\alpha \in A$, where $f_\alpha(x) = xK_\alpha$ for $x \in G$.

Suppose that $t \in T$ and $x \in G$; then

$$f_\alpha(txt^{-1}x^{-1}) = f_\alpha(t).\, f_\alpha(x).\, f_\alpha(t)^{-1}.\, f_\alpha(x)^{-1} = e.$$

Thus T is central in G. In fact, it is easily checked that T is the identity component of the centre of G. (For suppose that S is the identity component of the centre of G. Then $T \subseteq S$ since T is also connected in the centre of G. But $f_\alpha(S)$ is a connected closed subgroup of the centre of G_α and so $f_\alpha(S) \subseteq T_\alpha = f_\alpha(T)$. Thus $f_\alpha(T) = f_\alpha(S)$ and so 6.5.3 (ii) shows that $S = T$.) For each $\alpha \in A$,

$$f_\alpha(G) = G_\alpha = T_\alpha \Sigma_\alpha = f_\alpha(T) f_\alpha(\Sigma) = f_\alpha(T\Sigma)$$

and hence $G = T\Sigma K_\alpha$. Thus

$$G = \cap_\alpha T\Sigma K_\alpha = T\Sigma$$

by (6.5.4). Always $f_\alpha(T \cap \Sigma) \subseteq T_\alpha \cap \Sigma_\alpha$, the latter group being finite Abelian. Thus $T \cap \Sigma$ is a projective limit of finite Abelian groups and hence is totally disconnected Abelian (Exercise 6.F). Now we can apply A.1.6 with the result that:

<u>Fact 1.</u> The group G is topologically isomorphic with $(T \times \Sigma)/P$, where T and Σ are as above and P is a closed totally disconnected subgroup of the centre of $G \times \Sigma$. Further P and $T \cap \Sigma$ are isomorphic and $P \cap T = \{e\} = P \cap \Sigma$.

<u>Fact 2.</u> The group Σ is topologically isomorphic with a quotient $\Pi_{i \in I} G_i/Q$, where each G_i is a simple, simply connected, compact Lie group and Q is a totally disconnected subgroup of the centre of the product.

Let $p_\alpha : \tilde{\Sigma}_\alpha \to \Sigma_\alpha$ denote the simply connected covering group of Σ_α. Then Σ_α and $\tilde{\Sigma}_\alpha/Q_\alpha$ are isomorphic, where $Q_\alpha = \ker(p_\alpha)$ is a finite subgroup of the centre of $\tilde{\Sigma}_\alpha$. From 6.4.4 we know that $\tilde{\Sigma}_\alpha$ is compact. Suppose that $\beta > \alpha$ and let U_β be an open neighbourhood of the identity in $\tilde{\Sigma}_\beta$ such that $\tilde{\Sigma}_\alpha$ and Σ_α are locally isomorphic via p_α on $f_{\beta\alpha} \circ p_\beta(U_\beta)$. Denote the local inverse of p_α by q_α and define $\tilde{f}_{\beta\alpha} : U_\beta \to \tilde{\Sigma}_\alpha$ by

$$\tilde{f}_{\beta\alpha}(x) = q_\alpha \circ f_{\beta\alpha} \circ p_\beta(x).$$

Since $\tilde{\Sigma}_\beta$ is simply connected, $\tilde{f}_{\beta\alpha}$ has a unique extension to a homomorphism from $\tilde{\Sigma}_\beta$ into $\tilde{\Sigma}_\alpha$ (see the opening remarks of §3. 4). Let us denote this extension also by $\tilde{f}_{\beta\alpha}$. Then $\tilde{f}_{\beta\alpha}$ is a continuous open epimorphism. (In view of the open mapping theorem, all that needs to be checked is that $\tilde{f}_{\beta\alpha}$ is onto. Given $y \in \tilde{\Sigma}_\alpha$, since $\tilde{\Sigma}_\alpha$ is connected and $\tilde{f}_{\beta\alpha}(U_\beta)$ is an open neighbourhood of e in $\tilde{\Sigma}_\alpha$, we can choose $x_1, \ldots, x_n \in U_\beta$ such that

$$\tilde{f}_{\beta\alpha}(x_1 \ldots x_n) = \tilde{f}_{\beta\alpha}(x_1) \ldots \tilde{f}_{\beta\alpha}(x_n) = y.)$$

Evidently the following diagram commutes:

$$(6.\ 5.\ 7) \qquad \begin{array}{ccc} \tilde{\Sigma}_\beta & \xrightarrow{\tilde{f}_{\beta\alpha}} & \tilde{\Sigma}_\alpha \\ p_\beta \downarrow & & \downarrow p_\alpha \\ \Sigma_\beta & \xrightarrow{f_{\beta\alpha}} & \Sigma_\alpha \end{array} \qquad (\beta > \alpha)$$

It is easily seen that the set A, the groups $\tilde{\Sigma}_\alpha$ and the mappings $\tilde{f}_{\beta\alpha}$ form an inverse mapping system. Write $\tilde{\Sigma} = \varprojlim \tilde{\Sigma}_\alpha$. Given $\tilde{x} = (\tilde{x}_\alpha) \in \tilde{\Sigma}$, in view of (6. 5. 7), $(p_\alpha \tilde{x}_\alpha) \in \phi(\Sigma)$; define

$$p : \tilde{\Sigma} \to \tilde{\Sigma} \text{ by } p(\tilde{x}) = \phi^{-1}((p_\alpha \tilde{x}_\alpha)_{\alpha \in A}).$$

Then p is a continuous homomorphism. Now $p(\tilde{\Sigma})$ is a closed subgroup of Σ and so, by 6. 5. 3 (ii), $\phi \circ p(\tilde{\Sigma})$ is the projective limit of the groups $f_\alpha \circ p(\tilde{\Sigma})$. Given $y_\beta \in \Sigma_\beta$, there exists $\tilde{y}_\beta \in \tilde{\Sigma}_\beta$ with $p_\beta \tilde{y}_\beta = y_\beta$. Arguing as in 6. 5. 3 (iv), there exists $\tilde{x} = (\tilde{x}_\alpha) \in \varprojlim \tilde{\Sigma}_\alpha$ such that $\tilde{x}_\beta = \tilde{y}_\beta$. Hence $f_\beta \circ p(\tilde{x}) = p_\beta \tilde{x}_\beta = y_\beta$ and so $\phi \circ p(\tilde{\Sigma})$ is the projective limit of the Σ_α. Thus p is surjective. We now know that $p : \tilde{\Sigma} \to \Sigma$ is a continuous open epimorphism. Let Q denote its kernel; then

(6.5.8) Σ and $\tilde{\Sigma}/Q$ are topologically isomorphic.

Turning to Q, $\tilde{x} = (\tilde{x}_\alpha) \in Q$ if and only if $\tilde{x}_\alpha \in \ker p_\alpha = Q_\alpha$ for each $\alpha \in A$. Thus

$$f_\alpha(Q) \subseteq Q_\alpha \text{ for each } \alpha \in A$$

and so Q is a totally disconnected Abelian group. Also the centrality of each of the Q_α yields the same fact about Q.

From 6.4.5 and (6.5.6) we know that $\tilde{\Sigma}_\alpha$ is analytically isomorphic with $(H_{\alpha 1} \times \ldots \times H_{\alpha m(\alpha)})/K_\alpha$, where the $H_{\alpha j}$ are compact, simply connected, simple Lie groups with Lie algebras $L_{\alpha j}$ respectively, and where K_α is a finite subgroup of the centre of the product. (There is no torus since $\tilde{\Sigma}_\alpha$ is semisimple.) Since each $H_{\alpha j}$ is simply connected, so is their product. But this product is locally isomorphic with its quotient by K_α and since this quotient is also simply connected (it is equal to $\tilde{\Sigma}_\alpha$), it must be the case that K_α is trivial. Hence

$$\tilde{\Sigma}_\alpha \cong H_{\alpha 1} \times \ldots \times H_{\alpha m(\alpha)} .$$

If $\alpha < \beta$, then $(\tilde{f}_{\beta\alpha})_{*,e} : L_{\beta 1} \oplus \ldots \oplus L_{\beta m(\beta)} \to L_{\alpha 1} \oplus \ldots \oplus L_{\alpha m(\alpha)}$ satisfies the conditions of Lemma 6.5.4 and so by Remark 6.5.5 it can be considered as a projection from a finite sum onto a partial sum. Hence if we identify analytically isomorphic Lie groups,

$$\tilde{f}_{\beta\alpha} : H_{\beta 1} \times \ldots \times H_{\beta m(\beta)} \to H_{\alpha 1} \times \ldots \times H_{\alpha m(\alpha)}$$

is a projection. (Use the fact that simply connected Lie groups are isomorphic if and only if the same is true for their Lie algebras; Corollary 3.4.6.) More specifically, if $\tilde{f}_{\beta\alpha} : H_{\beta i} \to H_{\alpha j}$ is an analytic isomorphism, either replace $H_{\beta i}$ by $H_{\alpha j}$ or vice versa and replace the restriction of $\tilde{f}_{\beta\alpha}$ to $H_{\beta i}$ by the identity map. In view of (6.5.1) it is possible to repeat this process for all pairs (α, β) with $\alpha < \beta$ in a consistent manner. This done, Example 6.5.2 is now applicable and shows that $\tilde{\Sigma}$ is a direct product of compact, simply connected, simple Lie groups G_i, $i \in I$ (these groups are taken from the $H_{\alpha j}$, $\alpha \in A$ and $1 \leq j \leq m(\alpha)$).

This completes the proof of Fact 2.

Fact 3. Suppose that H is a closed, totally disconnected, normal subgroup of a compact group G, and that T is a subgroup of the same type of G/H. Then there exists a closed, totally disconnected subgroup K of G such that $(G/H)/T$ and G/K are topologically isomorphic.

Suppose that $\pi : G \to G/H$ is the natural projection and define $K = \pi^{-1}(T) \supseteq H$. Then H is a closed normal subgroup of K and topologically

$$T \cong K/H \quad \text{via} \quad xH \mapsto xH$$

for all $x \in K$. This fact combines with the second isomorphism theorem for topological groups ((5. 34) and (5. 35) of Hewitt and Ross [1]) to show that

$$(G/H)/T \cong (G/H)/(K/H) \cong G/K,$$

where the isomorphisms are topological. (Notice that K is closed and normal in G.)

Let U be the identity component of K. As in 6. 5. 3 (iii), $\pi(U)$ is the identity component in T which must be trivial since T is totally disconnected. Hence

$$U \subseteq \{x \in G : \pi(x) \in T \text{ and } \pi(x) = e\} \subseteq H.$$

But H is also totally disconnected; thus $U = \{e\}$ and so K is totally disconnected as required.

We can now complete the proof of 6. 5. 6. From facts 1 and 2,

$$G \cong (T \times (\Pi_{i \in I} G_i)/Q)/P.$$

Define $Q' = \{e_T\} \times Q$; then $T \times (\Pi_i G_i/Q)$ and $(T \times \Pi_i G_i)/Q'$ are topologically isomorphic. Now invoke fact 3 to see that

$$G \cong (T \times \Pi_{i \in I} G_i)/K,$$

where T is a compact connected Abelian group, $(G_i : i \in I)$ is a family of compact, simply connected, simple Lie groups and K is a closed, totally disconnected, normal (and hence central, by 6. 4. 1) subgroup of the product. Finally the fact that $K \cap T = \{e\}$ follows from the similar property for P expressed in fact 1. Also $K_n \Pi_i G_i \subseteq Q$. //

Notes

The result that a compact group is a Lie group if and only if it is a closed subgroup of some unitary group was proved in 1934 by Pontrjagin [2]. The other equivalent conditions given in §6. 1 for a compact group to be Lie are based on (28. 61) and (44. 45) of Hewitt and Ross [2] (although these authors make no explicit mention of Lie groups).

As mentioned in 6. 2. 1, it wasn't until 1935 that an example, due to Birkhoff [1], of a connected Lie group not isomorphic to any linear group was given. This problem was explicitly referred to in 1930 by Cartan [3, §26] but, understandably, seems to have been regarded as an important open question from a much earlier date.

In a number of problems in harmonic analysis on infinite compact groups it transpires that there are three main classes of groups to be considered and that each class needs its own special techniques. (For example, see Figà-Talamanca and Price [1, Lemma 3. 1(b)] and [2].) For actual applications, the most useful description of these classes is in terms of the duals of their members. This can be done so: each class is made up of all infinite compact groups G such that

(i) Γ is finitely generated;

(ii) Γ is not finitely generated but there exists $\gamma \in \Gamma$ such that $[\{\gamma\}]$ is infinite; and

(iii) $[\{\gamma\}]$ is finite for all $\gamma \in \Gamma$.

In view of 6. 1. 1 above and some standard results on duals of compact groups (see Hewitt and Ross [2, §28]) it follows that the above classes of groups have the following direct descriptions:

(i) Lie groups;

(ii) Non-Lie groups G which have a closed normal subgroup H such that G/H is Lie; and

(iii) 0-dimensional groups.

The idea of a simple Lie algebra (or rather a simple transformation group) was understood by Lie who described the four main sequences of simple Lie algebras mentioned in 6.4.6. Following this, W. K. J. Killing [1] showed that apart from these four sequences, there were only five more simple Lie algebras, the so-called exceptional algebras. Killing's analysis was based on the characteristic equation of the adjoint transformation

$$\text{ad } Y : X \to [X, Y],$$

the quadratic part of which was later to be called the 'Killing form'. Killing also introduced the notion of a semisimple Lie algebra as one that is a direct sum of simple algebras.

However, in 1894 Élie Cartan [1, p. 9] wrote in his dissertation:

> 'Malheureusement les recherches de M. Killing manquent de rigeur, et notament, en ci qui concerne les groupes qui ne sont pas simples, il fait constamment usage d'un théorème qu'il ne démontre pas dans sa généralité; je signale dans ce travail un exemple où ce théorème n'est pas vérifié, et, quand l'occasion s'en presente, un certain nombre d'autres erreurs de moindre importance ... Le présent travail a pour but d'exposer et de completer en certains points les recherches de M. Killing, en y introduisant toute la rigeur désirable.'

In this work Cartan developed the Killing form into a powerful tool and showed that a Lie algebra is semisimple if and only if its Killing form is nondegenerate and if and only if it is noncommutative with no solvable (hence commutative) ideals (Cartan [1, pp. 51-2]).

It is interesting to note that in contrast to the case of Lie groups, the class of finite simple groups seems to be far from being classified. By considering groups of automorphisms of simple Lie algebras over finite Galois fields of prime-power order, six sequences of finite simple groups can be obtained, these sequences being closely related to the classical sequences of Lie groups. Also each of the alternating groups A_k, $k \neq 5$, is simple. However, over the last decade many finite simple groups have been found which seem to bear no relationship with any of the above infinite sequences or with the exceptional Lie groups. Such finite simple groups are usually referred to as sporadic. For a survey of the results pertaining to simple groups related to Lie algebras see Carter [1], while for a later survey incorporating the numerous sporadic groups known at that time see Feit [1].

F. Schur showed that if G is a C^k-Lie group (that is, if G is a C^k-manifold and the group operations are C^k) for $k \geq 2$, then it is an (analytic) Lie group. A sketch of the methods involved is given in Montgomery and Zippin [2, §4.4], while details for $k \geq 3$ are given in Pontrjagin [1, Chapter 9]. Schur's result led Hilbert to ask in 1900 whether in fact it was enough to assume that G was a C^0-Lie group (that is, a locally Euclidean Lie group) for it necessarily to be a Lie group. This question forms the centre of the so-called fifth problem of Hilbert, being the fifth of a list of twenty-three problems announced by Hilbert at the Second International Congress of Mathematicians held in Paris in 1900. To help appreciate the depth of this problem it should be pointed out that to a large extent the development of the theory of topological groups in the first half of this century is due to attempts to solve this problem. However, before describing some of the highlights of its history we mention the structure results of É. Cartan for compact connected Lie groups (which, in any case, came chronologically before the first major breakthrough in Hilbert's problem).

These results, given as Theorems 6.4.2 and 6.4.5 above, derive from Cartan [2, pp. 10-11] where it is shown that every compact Lie group is locally isomorphic to a product of a compact semisimple Lie group and a finite dimensional torus. See also Cartan [3, §52] and Pontrjagin [1, Theorem 87].

Actually the proofs of the structure theorems for compact Lie groups as given above are based on a number of important results, two of which remain unproven here. The first, due to Weyl [1, Kap. IV, Satz 2], is Theorem 6.4.4 above and states that the (simply connected) covering group of a compact semisimple Lie group is itself compact while the second is Cartan's criterion for a Lie algebra to be semisimple (see 6.3.5 above). Cartan's criterion for semisimplicity was used to prove that all the derivations of a semisimple Lie algebra are inner (Corollary

6. 3. 8), which in turn was used to prove Theorem 6. 3. 11, the latter result having the consequence that Lie subalgebras of compact Lie algebras are compact. This latter result combined with Weyl's theorem to yield Theorem 6. 3. 13, this result being used in the proof of the structure theorem 6. 4. 2 to show that the G_j are closed and compact, hence that K is finite. Without Cartan's criterion and Weyl's theorem (without, in fact, any mention of covering groups and simple connectedness), the proof of 6. 4. 2 would yield: 'Every compact connected Lie group G is topologically isomorphic with

$$(T_0 \times G_1 \times \ldots \times G_m)/K,$$

where T_0 is the identity component of the centre of G, the G_j are all the simple normal connected Lie subgroups of G, and K is a discrete subgroup of the centre of the product. Furthermore, $T_0 \cap K = \{e\} = G_j \cap K$ for $j \in \{1, 2, \ldots, m\}$.'

Returning to Hilbert's fifth problem, in 1933 Haar proved the existence of an invariant measure on a compact group which satisfied the second axiom of countability and in the same year von Neumann [1] used this result to solve the problem in the affirmative for compact locally Euclidean groups. (At about the same time this answer was also given to Hilbert's fifth problem for the case of Abelian locally Euclidean groups by Pontrjagin. It appears as Theorem 44 of Pontrjagin [1].) So it was only after thirty-three years that the first general case of this problem was settled, but nevertheless almost another two decades passed before the problem was solved in its full generality ... in 1952 Gleason [1] and Montgomery and Zippin [1] solved it affirmatively for all locally Euclidean groups.

Their proof involved the dimension of an arbitrary locally compact group, a notion which generalizes the dimension of a Lie group. (For the definition, see Montgomery and Zippin [2, p. 176].) Gleason showed that every finite dimensional locally compact group without small subgroups is a Lie group while Montgomery and Zippin proved the non-existence of small subgroups in locally connected groups of finite dimension. Together these results constituted an affirmative solution to

Hilbert's fifth problem since locally Euclidean groups are both finite dimensional and locally connected.

In 1946, Chevalley [1, p. 193] noted that a Lie group has no small subgroups. (See 2.2.14 above.) This observation led to the conjecture that every locally compact group with no small subgroups must be a Lie group. It was quickly realized that this conjecture was closely related to Hilbert's fifth problem for as Gleason [1, p. 193] wrote:

'The possible existence of small subgroups in a locally euclidean group has long been recognized as one of the chief difficulties involved in Hilbert's fifth problem.'

Less than a year after Gleason, Montgomery and Zippin had submitted their solution of Hilbert's fifth problem to the editors of the Annals of Mathematics, H. Yamabe [1] solved the latter conjecture in 1953, also in the affirmative. Details of all these results can be found in Montgomery and Zippin [2].

Returning to the case of compact groups, in §6.1 it was seen that the fact that such groups are Lie whenever they have no small subgroups is a simple consequence of the Peter-Weyl theorem. However, before a solution of Hilbert's fifth problem for compact groups can be deduced from the results in Chapter 6, the notion of dimension must be introduced into the structure theorem 6.5.6 for compact connected groups.

The essentials of this structure theorem are attributed to Pontrjagin [2] by Weil [1, p. 93] although it is apparent that Weil had to do a considerable amount of 'reading between the lines' to arrive at his presentation. The proof given above follows in outline that of Weil [1, §25]. By keeping careful track of the dimensions of the groups involved, it can be shown that

$$\dim(G) = \dim(T) + \sum_{i \in I} \dim(G_i),$$

where $\dim(\ )$ denotes the (possibly infinite-valued) dimension function and G is a compact connected group decomposed as in 6.5.6. If G is also locally connected and finite dimensional, then I is finite, T is a finite dimensional torus and K is finite; hence G is Lie. (For further details, see Weil [1, §25].)

Because so much more is known of the structure and representation theory of compact simple groups than of compact groups in general, it is frequently the case that problems on compact groups can be approached more easily via this subclass. However, the piecing together of the solutions on these component 'simple' parts via 6. 5. 6 often allows a full solution to be obtained. For examples of this procedure to various problems in harmonic analysis, see Armstrong [1], Cecchini [1] and Rider [1]. Recently Y-K. Yu [1] has used 6. 5. 6 to characterize certain types of topologically simple groups.

In 6. 4. 3 we came to the conclusion that a compact connected Lie group is semisimple if and only if it has a finite centre. An alternative description can be given in terms of a simple property of its representations introduced in McMullen and Price [1], namely that of 'tallness'. A compact group is said to be <u>tall</u> if it has at most a finite number of pairwise inequivalent, irreducible, continuous representations of any given degree. Extending a result of D. Rider, Hutchinson [1] has recently shown that, amongst other characterizations, a compact Lie group is semisimple if and only if it is tall. The proof uses a corollary of the structure theorem 6. 4. 2 above. The notion of whether or not a compact group is tall is another dichotomy similar to the trichotomy discussed in the third paragraph of these notes in that sometimes the different subclasses require quite different methods of proof. For example, see McMullen and Price [1].

Exercises

6. A. Show that every complex matrix of prime order p, $p > 1$, has an eigenvalue which is a primitive p-th root of unity.

6. B. Show directly that the group G_3^* defined after the proof of 6. 2. 2 is a connected Lie group of dimension 3.

6. C Disconnected non-linear Lie groups. (i) A group is said to be <u>torsion group</u> if every element is of finite order. Show that every Abelian algebraic torsion subgroup of $GL(n, C)$ is isomorphic to a subgroup of $(Q/Z)^n$, where Q is the additive group of rational numbers and Z is the additive group of integers.

(ii) Show that the product of an infinite number of copies of Q/Z is not an algebraic subgroup of any GL(n, C). In fact, show that an infinite product of copies of Z_2 is not an algebraic subgroup of any GL(n, C).

(iii) If we temporarily remove the caveat against 0-dimensional spaces imposed in 1.1.1, then a topological group is discrete if and only if it is a 0-dimensional Lie group. Thus the products described in (ii) above are examples of 0-dimensional Lie groups which are not linear. Use these examples to construct Lie groups of arbitrary order which are not linear.

6.D. Let f be a symmetric bilinear form on a real finite-dimensional space E and let F be a subspace of E. If $F^\perp = \{x \in E : f(x, y) = 0 \text{ for all } y \in F\}$, then the following three conditions are equivalent:

(i) the restriction of f to F is nondegenerate;

(ii) $F \cap F^\perp = \{0\}$;

(iii) $F \cap F^\perp = \{0\}$ and $F \oplus F^\perp = E$.

If f is nondegenerate on E, then these conditions are also equivalent to:

(iv) the restriction of f to $F^\perp$ is nondegenerate.

6.E **Essential uniqueness of the Killing form.** The following exercises involve complex Lie algebras; the definitions remain the same as their real counterparts except that R is replaced throughout by C. Furthermore, Cartan's criterion for semisimplicity remains intact for the complex case. The format of the questions is based on Sagle and Walde [1, p. 242].

(a) Suppose that L is a simple complex Lie algebra and that f is a nondegenerate bilinear form on L which is invariant (that is,

$$f([X, Y], Z) = f(X, [Y, Z])$$

for all X, Y, Z ∈ L). The following steps form a proof of the fact that

$f(x, Y) = \lambda K(X, Y)$

for all $X, Y \in L$ and some $\lambda \in C$.

(i) Show that there exists $T \in GL(L)$ such that

$$f(X, Y) = K(TX, Y)$$

for all $X, Y \in L$.

(ii) By considering $K([\text{ad } X, T]Y, Z)]$, where $X, Y, Z \in L$, show that $[\text{ad } X, T] = 0$ for all $X \in L$.

(iii) Use Schur's lemma (see A. 3. 7) to show that $T = \lambda I$ for some $\lambda \in C$.

(b) When L is a complex simple Lie subalgebra of $\underline{\underline{gl}}(n, C)$ then there exists $\lambda \in C$ such that

$$K(X, Y) = \lambda \operatorname{tr}(XY)$$

for $X, Y \in L$. This can be proved via the following steps.

(i) Show that $(X, Y) \mapsto \operatorname{tr}(XY)$ is an invariant bilinear form on L.

(ii) Define $L^\perp = \{X \in L : \operatorname{tr}(XY) = 0 \text{ for all } Y \in L\}$. Show that $L^\perp = L$ or $\{0\}$.

(iii) If M is a Lie algebra, define $M^{(1)}$ to be the set in M generated by $\{[X, Y] : X, Y \in M\}$. Show that $M^{(1)}$ is an ideal of M. By induction, define $M^{(n+1)}$ to be the ideal of $M^{(n)}$, and hence of M, generated by $\{[X, Y] : X, Y \in M^{(n)}\}$. If there exists n such that $M^{(n)} = \{0\}$, then M is said to be <u>soluble.</u> A deep result due to Cartan states that if M is a Lie subalgebra of $\underline{\underline{gl}}(n, C)$, then M is soluble if $\operatorname{tr}(XY) = 0$ for all $X, Y \in M$. Use this to show that $L^\perp = L$ is an impossibility. Now complete the proof of (b).

6. **F.** (i) Show that a topological group is a projective limit of finite groups (a <u>profinite group)</u> if and only if it is compact and totally disconnected.

(ii) A compact group G is connected and Abelian if and only if it can be written as a projective limit of torii. (See 6. 5. 3 (v).)

6. G **Projective limits.** Suppose that the collection consisting of a directed set A, topological groups G_α $(\alpha \in A)$ and continuous homomorphisms $f_{\beta\alpha} : G_\beta \to G_\alpha$ $(\alpha < \beta)$ is an inverse mapping system.

(i) Show that the projective limit $G = \varprojlim G_\alpha$ defined in 6.5.1 has the properties:

(a) for each $\alpha \in A$ there exists a continuous homomorphism $f_\alpha : G \to G_\alpha$ such that $f_\alpha = f_{\beta\alpha} \circ f_\beta$ for all $\alpha < \beta$;

(b) given a topological group H and continuous homomorphisms $h_\alpha : H \to G_\alpha$ such that $h_\alpha = f_{\beta\alpha} \circ h_\beta$ for all $\alpha < \beta$, then there exists a unique continuous homomorphism $h : H \to G$ such that $h_\alpha = f_\alpha \circ h$ for all $\alpha \in A$, that is, such that the diagram

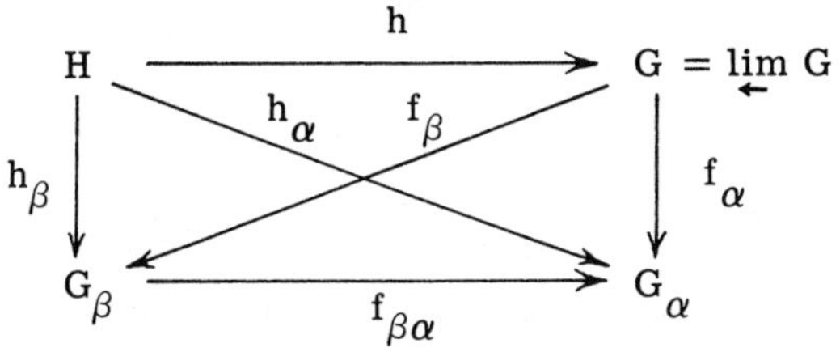

commutes whenever $\alpha < \beta$.

(ii) Show that, up to isomorphism, $G = \varprojlim G_\alpha$ is the only topological group satisfying properties (a) and (b). (Since this characterizes projective limits via a universal property, it opens the way to introduce this type of limit in more general categories. See MacLane [1, p. 68].)

Appendix · Abstract harmonic analysis

In this appendix a few of the basic results in the vast theory of locally compact groups and their representations are given. These results will be stated without proofs; for these the interested reader is recommended to consult the major work of Hewitt and Ross [1], [2] for results in general, or the lecture notes of Edwards [2] for results pertaining to harmonic analysis on compact groups.

A.1 Topological groups

A.1.1 Definition. A group G which is also a Hausdorff topological space is said to be a topological group if both the maps

$(x, y) \mapsto xy$ from $G \times G$ into G, and

$x \mapsto x^{-1}$ from G into G

are continuous.

The identity element of a group will always be denoted by e, and all topologies will be assumed to be Hausdorff.

A.1.2 Theorem. Let G be a topological group and let $\mathfrak{U}$ be an open basis at e. Then:

(i) $\cap \{U : U \in \mathfrak{U}\} = \{e\}$;

(ii) for every $U \in \mathfrak{U}$, there is $V \in \mathfrak{U}$ such that $V^2 \subseteq U$;

(iii) for every $U \in \mathfrak{U}$, there is $V \in \mathfrak{U}$ such that $V^{-1} \subseteq U$;

(iv) for every $U \in \mathfrak{U}$ and $x \in U$, there is $V \in \mathfrak{U}$ such that $xV \subseteq U$;

(v) for every $U \in \mathfrak{U}$ and $x \in G$, there is $V \in \mathfrak{U}$ such that $xVx^{-1} \subseteq U$.

Conversely, let G be a group and let $\mathfrak{U}$ be a family of subsets of G each containing e which satisfies conditions (i)-(v) above. Then the family of sets xU, where x runs through G and U runs through $\mathfrak{U}$, is an open subbase for a topology on G. Under this topology G is a topological group. (The family of sets Ux is a subbase for the same topology.)

Clearly every topological group has an open basis at e consisting of symmetric sets, that is, consisting of sets U such that $U = U^{-1}$.

A. 1. 3 **Definition.** A topological group F is said to be <u>locally compact</u> if every point of G has a compact neighbourhood.

If G is a locally compact topological group, then it possesses an open basis about e consisting entirely of relatively compact, symmetric sets.

A. 1. 4 **Products of groups.** Suppose that we are given a topological group G_α for each member of α of an index set A. (Briefly we say that $(G_\alpha : \alpha \in A)$ is a family of topological groups.) Then $\Pi_{\alpha \in A} G_\alpha$ is said to be the direct product of the G_αs and is defined as their Cartesian product equipped with the product topology and the group product $(x_\alpha)(y_\alpha) = (x_\alpha y_\alpha)$. The direct product of a family of (compact) topological groups is a (compact) topological group. This result is also valid for locally compact groups when the family is finite.

A. 1. 5 **Open mapping theorem.** Let G be a σ-compact locally compact group and H a locally compact group. Then every continuous epimorphism from G onto H is open.

A. 1. 6 **Corollary.** <u>Suppose that</u> G <u>is a compact group and that</u> G_1, G_2 <u>are closed subgroups of</u> G <u>which commute (that is,</u> $xy = yx$ <u>whenever</u> $x \in G_1$ <u>and</u> $y \in G_2$). <u>If</u> $G_1 G_2 = G$, <u>then there exists a closed normal subgroup</u> K <u>of</u> $G_1 \times G_2$ <u>such that</u>

(i) K <u>is topologically isomorphic with</u> $G_1 \cap G_2$;

(ii) $K \cap (G_1 \times \{e_2\}) = \{e\} = K \cap (\{e_1\} \times G_2)$; <u>and</u>

(iii) G <u>and</u> $(G_1 \times G_2)/K$ <u>are topologically isomorphic.</u>

Proof. The map $f : G_1 \times G_2 \to G$ given by $f(x, y) = xy$ is clearly a continuous epimorphism. It is open by the proceding theorem. Let $K = \ker(f)$; then G and $(G_1 \times G_2)/K$ are topologically isomorphic (Theorem (5.34) of Hewitt and Ross [1]). It remains to settle (i) and (ii).

If both (x, y_1) and $(x, y_2) \in K$, then

$$xy_1 = xy_2 = e;$$

thus $y_1 = y_2 = x^{-1}$. Hence

$$K = \{(x, x^{-1}) : x \in G_1, x^{-1} \in G_2\} = \{(x, x^{-1}) : x \in G_1 \cap G_2\}$$

and so the map $\psi : (x, x^{-1}) \mapsto x$ provides the required topological isomorphism between K and $G_1 \cap G_2$. Moving on to (ii), if (x, x^{-1}) is a typical element of K (with $x \in G_1 \cap G_2$), then $(x, x^{-1}) \in G_1 \times \{e_2\}$ if and only if $x = e_1$. //

A. 2 Representations

Let E be a complex Hilbert space and let $GL(E)$ denote the space of invertible linear bicontinuous operators on E. A map γ from a group G into $GL(E)$ is said to be a representation of G provided

$$\gamma(xy) = \gamma(x)\gamma(y) \text{ for all } x, y \text{ in } G.$$

The dimension of the representation γ is defined as the dimension of E. (To avoid trivialities we suppose that E is of dimension at least 1.) Unitary representations are representations whose ranges are contained in the set of unitary maps in $GL(E)$. For example, the identity representation $\gamma : G \to GL(E)$, where $\gamma(x) = I$ for all x, is unitary.

Suppose that G is a topological group. Then a representation γ is said to be weakly continuous, strongly continuous or uniformly continuous according as γ is continuous when $GL(E)$ is equipped with the weak operator topology, the strong operator topology or the uniform operator topology. (See Naimark [1, §33].) These types of continuity of γ are equivalent to the continuity of

$x \mapsto \langle \gamma(x)v, w\rangle$ from G into C for each v, w in E,

$x \mapsto \gamma(x)v$ from G into E for each v in E, and

$x \mapsto \gamma(x)$ from G into GL(E), where GL(E) is equipped with the usual operator norm, respectively.

A. 2. 1 Theorem. Suppose that G is a topological group, E is a Hilbert space, and $\gamma : G \to GL(E)$ is a representation.

(i) If G is also locally compact and γ is unitary, then γ is weakly continuous if and only if it is strongly continuous.

(ii) If E is finite-dimensional, then the three types of continuity are equivalent.

The proof of part (i) follows from Hewitt and Ross [1, (22.8)], while that of part (ii) is an immediate consequence of the fact that all Hausdorff topologies on a finite-dimensional linear space which make it into a topological vector space are equivalent (Edwards [1, Proposition 1.9.6]).

In view of this result we say that a representation is continuous if it is strongly continuous.

A representation $\gamma : G \to GL(E)$ is said to be reducible if there exists a closed subspace F of E such that $\{0\} \neq F \neq E$ and $\gamma(x)F \subseteq F$ for all x in G; otherwise γ is said to be irreducible.

A. 2. 2 Theorem (Gel'fand-Raikov). Let G be a locally compact group. For every x in G with $x \neq e$ there is a continuous, irreducible, unitary representation γ of G such that $\gamma(x) \neq I$.

Two representations $\gamma_1 : G \to GL(E_1)$ and $\gamma_2 : G \to GL(E_2)$ are said to be equivalent if there exists an isometric isomorphism W from E_1 onto E_2 such that $W\gamma_1(x) = \gamma_2(x)W$ for all x in G.

A. 2. 3 Direct sums of representations. Suppose that we are given a family of representations $(\gamma_\alpha : \alpha \in A)$. Define $E = \oplus_{\alpha \in A} E_\alpha$, the Hilbert-space direct sum of $(E_\alpha : \alpha \in A)$. (Each element of E is of the form $v = \oplus_\alpha v_\alpha$, where $\sum_\alpha \|v_\alpha\|^2 < \infty$, and the inner product on E is $\langle v, w\rangle = \sum_\alpha \langle v_\alpha, w_\alpha\rangle$.) Whenever

$$\text{(A. 2. 1)} \quad \sup\{\|\gamma_\alpha(x)\| : \alpha \in A,\ x \in G\} < \infty,$$

then we define the direct sum of the family (γ_α) as the representation $\gamma : G \to GL(E)$ where

$$\gamma(x)v = \oplus_{\alpha \in A}\gamma_\alpha(x)v_\alpha$$

for each x in G and $v = \oplus_\alpha v_\alpha$ in E. If (A. 2. 1) is satisfied then clearly γ is (continuous) unitary if and only the same is true of each γ_α. Also notice that (A. 2. 1) is automatically satisfied when each of the γ_α is unitary.

(i) Suppose that γ is a unitary representation; then it can be written as the direct sum of two nontrivial unitary representations if and only if it is reducible. (The reducibility of $\gamma : G \to GL(E)$ implies precisely that there exists a closed subspace F of E such that $\{0\} \neq F \neq E$ and $\gamma(x)F \subseteq F$. The orthogonal complement $F^\perp$ of F also satisfies $\gamma(x)F^\perp \subseteq F^\perp$. For each x in G define $\gamma_1(x) = \gamma(x)|_F$ and $\gamma_2(x) = \gamma(x)|_{F^\perp}$; then $\gamma_1 : G \to GL(F)$ and $\gamma_2 : G \to GL(F^\perp)$ are unitary representations and furthermore $\gamma = \gamma_1 \oplus \gamma_2$. The converse is also trivial.)

(ii) Every finite-dimensional unitary representation is the direct sum of a finite number of irreducible unitary representations.

A. 2. 4 **Abelian groups.** Every continuous, unitary, irreducible representation of an Abelian group is 1-dimensional.

A. 2. 5 **G-spaces.** If $\gamma : G \to GL(E)$ is a continuous representation, then it is sometimes convenient to suppress explicit mention of the homomorphism γ and simply say that E is a G-space or that G acts on E. Thus, in analogy with representations, we have the notions of irreducible, unitary G-spaces and of direct sums of G-spaces.

A. 3 Compact groups

One of the reasons why the study of harmonic analysis becomes relatively tractable when we pass from general locally compact groups to compact groups is:

A. 3. 1 **Theorem.** All irreducible, continuous, unitary representations of compact groups are finite-dimensional.

It follows from A. 2. 3 (ii) that every continuous unitary, finite-dimensional representation of a compact group can be written as a direct sum of a finite number of irreducible such representations. In fact, more is true ... every such decomposition is unique up to order and equivalences.

A. 3. 2 **Dual objects.** When considering the collection of all continuous, irreducible, unitary representations of a compact group G as a whole, it clearly serves little purpose to include a representation along with all the other representations equivalent to it. To overcome this, select one representation from each equivalence class of such representations, this selection to be made once for all. The set of these selections will be called the dual object, or dual, of G and will be denoted by Γ (or by $\Gamma(G)$ whenever explicit mention of the underlying group is necessary).

A. 3. 3 **Conjugates and tensor products.** Let E be a Hilbert space. Then there is a conjugate linear function D carrying E onto itself such that

(i) $\langle Dv, Dw\rangle = \langle v, w\rangle$, and

(ii) $D^2 = I$.

Let $\gamma \in \Gamma$; then $x \mapsto D\gamma(x)D$ is also a continuous, irreducible, unitary representation of G and we let $\bar{\gamma}$ denote its representative in Γ. The representation $\bar{\gamma}$ is called the conjugate of γ.

Even though we may have started with a different conjugate-linear map satisfying (i) and (ii), the resultant $\bar{\gamma}$ is still the same. One way to see this is to first choose an orthonormal basis $v_1, \ldots, v_n$ of E, where G acts on E via γ; then for each x in G, $\gamma(x)$ is represented by an $n \times n$ matrix $(\gamma_{ij}(x))$. In terms of this representation $\bar{\gamma}$ is always equivalent to the representation $x \mapsto \overline{(\gamma_{ij}(x))}$.

Whenever E and F are finite-dimensional Hilbert spaces we denote their tensor product by $E \otimes F$ ... $E \otimes F$ is also a finite-dimensional space and as such it can be given the structure of a Hilbert space in a natural manner. If $S \in GL(E)$ and $T \in GL(F)$, then there exists a unique operator in $GL(E \otimes F)$ which maps $u \otimes v$ into $Su \otimes Tv$. Denote this operator by $S \otimes T$. The _tensor product_ of $\beta, \gamma \in \Gamma$, where $\beta : G \to GL(E)$ and $\gamma : G \to GL(F)$, is defined as the representation

$$\beta \otimes \gamma : x \mapsto \beta(x) \otimes \gamma(x)$$

from G into $GL(E \otimes F)$. It is readily checked that $\beta \otimes \gamma$ is a continuous unitary representation. However it is not necessarily irreducible.

Let G be a compact group and let Δ be a subset of Γ. Then $[\Delta]$ will denote the smallest subset of Γ which is closed under the operations of

(i) conjugation, and

(ii) tensor products followed by decomposition into irreducible direct summands in Γ.

In this case Γ is said to be finitely-generated if $\Gamma = [\Delta]$ for some finite set Δ. (When Γ is equipped with the operators (i) and (ii) defined immediately above then it has the algebraic structure of a hypergroup and as such it is often referred to as _the dual hypergroup of_ G.)

A. 3. 4 Theorem. _Let_ G _be a compact group and let_ Δ _be a subset of_ Γ. _Then_ $[\Delta] = \Gamma$ _if and only if for each_ x _in_ G, $x \neq e$, _there exists_ γ _in_ Δ _such that_ $\gamma(x) \neq I$. _(Cf. A. 2. 2.)_

A. 3. 5 Unitary representations. Nothing is gained by studying continuous, finite-dimensional representations of compact groups which are not unitary, since every such representation of a compact group is equivalent to one that is also unitary.

A. 3. 6 Trigonometric polynomials. Let $C(G, C)$ denote the set of continuous complex-valued functions on G, a compact group. Any function f in $C(G, \mathbf{C})$ which can be written in the form

$$f = \sum_{\gamma \in F} \operatorname{tr}[A(\gamma)\gamma(\cdot)],$$

where: F is a finite subset of Γ; tr denotes the usual trace of a finite-dimensional linear operator (or matrix); and for each γ in F, $\gamma : G \to GL(E_\gamma)$ is a member of Γ, and $A(\gamma)$ is an endomorphism of E_γ, is called trigonometric polynomial. An important result in the foundations of abstract harmonic analysis on compact groups is that the linear space of trigonometric polynomials is dense in many of the standard spaces of functions. For example, the trigonometric polynomials are dense in $C(G, \mathbf{C})$.

If f is also central (that is, if $f(xyx^{-1}) = f(y)$ for all x, y in G), then f can be uniformly approximated by central trigonometric polynomials. All central trigonometric polynomials are of the form $\sum_{\gamma \in F} c(\gamma)\chi_\gamma$, where F is a finite subset of Γ, and for each γ in F, $c(\gamma)$ is a complex number and χ_γ the central function $\chi_\gamma : x \mapsto \operatorname{tr}[\gamma(x)]$.

A. 3. 7 Corollary to Schur's lemma. Let E be a finite-dimensional space over $\mathbf{C}$ and let $\mathfrak{M}$ be an irreducible set of linear operators on E. (Here irreducible means that no proper linear subspace of E is invariant under all members of $\mathfrak{M}$.) Whenever A is a linear operator on E with $AM = MA$ for all $M \in \mathfrak{M}$, then A must be a scalar multiple of the identity operator.

A. 4 The Haar integral

A. 4. 1 **Basic definitions.** Let X be a locally compact Hausdorff space and let $C_c(X)$ denote the linear space of real-valued continuous functions on X with compact support. A positive integral on X is a linear functional μ on $C_c(X)$ for which $\mu(f) \geq 0$ whenever $f \geq 0$. The number $\mu(f)$ is called the integral of f over X with respect to μ and is frequently denoted by $\int_X f(x)d\mu(x)$ or simply $\int_X f\, d\mu$.

In the particular case when X is a locally compact group, G say, then a nontrivial positive integral μ on G is said to be a Haar integral if it is invariant under left translations, that is, if

$$\int_G f d\mu = \int_G \tau_a f d\mu$$

for all $f \in C_c(G)$ and $a \in G$, where $\tau_a f$ is the function $x \mapsto f(a^{-1}x)$.

A. 4. 2 **Theorem.** _On every locally compact group_ G _there exists at least one Haar integral_ μ. _Moreover, such an integral is unique in the sense that if_ ν _is another Haar integral on_ G, _then there exists a real number_ $c > 0$ _with_ $\nu = c\mu$.

A. 4. 3 **The compact case.** When G is a compact group, every Haar integral is also invariant under right translations. Furthermore, in this case it is customary to choose the unique Haar integral μ with the property that $\int_G 1 d\mu = 1$.

A. 4. 4 **Remarks.** Further details of the above approach to integration theory can be found in Edwards [1, Chapter 4]. It should be pointed out that the more usual approach to integration theory is via the notion of measures, that is, of countably additive, extended real-valued non-negative functions defined on suitable algebras of subsets of X. For details, see Hewitt and Ross [1, Chapter 3]. In practice these two approaches are essentially equivalent for locally compact spaces, a fact repeatedly testified to by the various versions of the Riesz representation theorem. For example, there is a canonical bijection between Haar integrals on a locally compact group G and non-negative regular Borel measures m on G which have the property that $m(E) = m(a^{-1}E)$ for all $a \in G$ and Borel subsets E of G.

Bibliography

Alan L. Armstrong. [1] On the derived algebra of L_p of a compact group. Michigan Math. J. 21 (1974), 341-52.

V. I. Averbukh and O. G. Smolyanov. [1] The various definitions of the derivative in linear topological spaces, Russian Math. Surveys 23 (1968), 67-113.

H. F. Baker. [1] Alternants and continuous groups, Proc. London Math. Soc. 3 (1905), 24-47.

Garrett Birkhoff. [1] Lie groups simply isomorphic with no linear group, Bull. Amer. Math. Soc. 42 (1936), 883-8.

N. Bourbaki. [1] Groupes et algèbres de Lie (Éléments de Mathématique, Chapters 2-3 (1972), Chapters 4-6 (1968), Hermann, Paris).

J. E. Campbell. [1] On a law of combination of operators, Proc. London Math. Soc. 29 (1898), 14-32.

Élie Cartan. [1] Sur la structure des groupes de transformations finis et continus (Thèse, 1894, Paris, Nony; 2e edition, Vuibert, 1933).

[2] Groupes simples clos et ouverts et géométrie riemannienne, J. Math. Pures Appl. 8 (1929), 1-33.

[3] La théorie des groupes finis et continus et l'analysis situs, Mémor. Sci. Math. 42 (1930).

R. W. Carter. [1] Simple groups and simple Lie algebras, J. London Math. Soc. 40 (1965), 193-240.

Carlo Cecchini. [1] Lacunary Fourier series on compact Lie groups, J. Functional Analysis 11 (1972), 191-203.

Claude Chevalley. [1] Theory of Lie groups, Volume 1 (Princeton University Press, Princeton, 1946).

E. A. Coddington and N. Levinson. [1] Theory of ordinary differential equations, (McGraw-Hill, New York, Toronto, London, 1955).

P. M. Cohn. [1] Lie groups, (Cambridge Tracts in Mathematics and Mathematical Physics, No. 46, C. U. P., Cambridge, 1957).

J. Dieudonné. [1] Foundations of modern analysis, (Academic Press, New York, 1960).

J. Dixmier and A. Douady. [1] Champs continus d'espaces hilbertiens et de C*-algèbres, Bull. Soc. Math. France 91 (1963), 227-84.

James Dugundji. [1] Topology (Allyn and Bacon, Boston, 1966).

N. Dunford and J. T. Schwartz. [1] Linear operators, Part 1: General theory, (Interscience, New York, London, 1958).

R. E. Edwards. [1] Functional analysis: Theory and applications, (Holt, Rinehart and Winston, New York, 1965).

[2] Integration and Harmonic analysis on compact groups, (London Mathematical Society Lecture Notes, Volume 8, Cambridge University Press, Cambridge, 1972).

James Eells, Jr. [1] A setting for global analysis, Bull. Amer. Math. Soc. 72 (1966), 751-807.

J. Eells and K. D. Elworthy. [1] Open embeddings of certain Banach manifolds, Ann. of Math. 91 (1970), 465-85.

Walter Feit. [1] The current situation in the theory of finite simple groups, Actes Congrès Intern. Math., Tome 1 (1970), 55-93 (Gauthier-Villars, Paris, 1971).

A. Figà-Talamanca and J. F. Price. [1] Applications of random Fourier series over compact groups to Fourier multipliers, Pacific J. Math. 43 (1972), 531-41.

[2] Rudin Shapiro sequences on compact groups, Bull. Austral. Math. Soc. 8 (1973), 241-5.

A. Gleason. [1] Groups without small subgroups, Ann. of Math. 56 (1952), 193-212.

F. Hausdorff. [1] Die symbolische Exponentialformel in der Gruppentheorie, Berichte der Sächsischen Akad. Wiss. (Math.-Phys. Kl.), Leipzig 58 (1906), 19-48.

S. Helgason. [1] Differential geometry and symmetric spaces, (Academic Press, New York, London, 1962).

David W. Henderson. [1] Infinite-dimensional manifolds are open subsets of Hilbert space, Topology 9 (1970), 25-33.

Edwin Hewitt and Kenneth A. Ross. [1] Abstract harmonic analysis: Volume 1 (Die Grundlehren der mathematischen Wissenschaften, Band 115. Springer-Verlag, Berlin, Heidelberg, New York, 1963).

[2] Abstract harmonic analysis: Volume 2 (as for Volume 1, Band 152, 1970).

John Horvath. [1] Topological vector spaces and distributions, (Addison-Wesley, Reading, London, Don Mills, 1966).

M. F. Hutchinson. [1] Non-tall compact groups admit infinite Sidon sets, J. Austral. Math. Soc. (to appear).

C. Itzykson and M. Nauenberg. [1] Unitary groups: representations and decompositions, Rev. Modern Phys. 38 (1966), 404-28.

Nathan Jacobson. [1] Lie algebras, (Interscience, New York, London, 1962).

Robert R. Kallman. [1] The topology of compact simple Lie groups is essentially unique, Advances in Math. 12 (1974), 416-7.

John L. Kelley. [1] General topology, (Van Nostrand, Princeton, Toronto, London, New York, 1955).

Wilhelm K. J. Killing. [1] Die Zusammensetzung der stetigen endlichen Transformationsgruppen, I Math. Ann. 31 (1888), 252-90; II ibid. 33 (1889), 1-48; III ibid. 34 (1889), 57-122; IV ibid. 36 (1890), 161-89.

Shoshichi Kobayashi and Katsumi Nomizu. [1] Foundations of differential geometry: Volume 1, (Interscience, New York, London, 1963).

Serge Lang. [1] Analysis II, (Addison-Wesley, Reading, London, Don Mills, Ontario, 1969).

[2] Differentiable manifolds, (Addison-Wesley, Reading, Don Mills, London, 1972).

S. Lie. [1] Theorie der Transformationsgruppen, Math. Ann. 16 (1880), 441-528; a translation appears in Sophus Lie's 1880 transformation group paper, translated by M. Ackerman, comments by R. Hermann (Math. Sci. Press, Massachusetts, 1975).

Saunders MacLane. [1] Categories for the working mathematician, (Springer-Verlag, New York, Heidelberg, Berlin, 1971).

Wilhelm Magnus, Abraham Karass and Donald Solitar. [1] Combinatorial group theory: Presentations of groups in terms of generators and relations, (Interscience, New York, London, Sydney, 1966).

W. Mayer and T. W. Thomas. [1] Foundations of the theory of Lie groups, Ann. of Math. 36 (1935), 770-822.

J. R. McMullen and J. F. Price. [1] Rudin-Shapiro sequences for arbitrary compact groups, J. Austral. Math. Soc. (to appear).

B. S. Mityagini. [1] The homotopy structure of the linear group of a Banach space, Russian Math. Surveys 25 (5) (1970), 59-103.

D. Montgomery and L. Zippin. [1] Small subgroups of finite-dimensional groups, Ann. of Math. 56 (1952), 213-41.

[2] Topological transformation groups, (Interscience, New York, London, 1955).

M. A. Naimark. [1] Normed rings, (Wolters-Noordhoff, Groningen, 1970).

L. S. Pontrjagin. [1] Topological groups, (Princeton Univ. Press, Princeton, 1939).

[2] Sur les groupes topologique compacts et le cinquième problème de M. Hilbert, C. R. Acad. Scie. Paris 198 (1934), 238-40.

D. Rider. [1] Central lacunary sets, Monatsh. Math. 76 (1972), 328-38.

A. Robinson. [1] Non-standard analysis, (North-Holland Publishing Company, Amsterdam, 1966).

Arthur A. Sagle and Ralph E. Walde. [1] Introduction to Lie groups and Lie algebras, (Academic Press, New York, London, 1973).

Hans Samelson. [1] Notes on Lie algebras, (Van Nostrand-Reinhold, New York, Cincinnati, Toronto, London, Melbourne, 1969).

Norman J. Steenrod. [1] The topology of fibre bundles, (Princeton University Press, Princeton, 1951).

Michael Spivak. [1] Differential geometry, Volume 1, (Michael Spivak, Brandeis University, 1970).

J. von Neumann. [1] Die Einführung analytischer Parameter in topologischen Gruppen, Ann. Math. 34 (1933), 170-90.

A. Weil. [1] L'integration dans les groupes topologiques et ses applications (Hermann, Paris, 1940).

H. Weyl. [1] Theorie der Darstellung kontinuierlicher halbeinfacher Gruppen durch lineare Transformationen: I, II, III und Nachtrag, Math. Z. 23 (1925), 271-309; 24 (1926), 328-76; 377-95; 789-91.

H. Yamabe. [1] A generalization of a theorem of Gleason, Ann. Math. 58 (1953), 351-65.

Ying-King Yu. [1] Topologically semisimple groups.

M. Hausner and J. Schwartz. [1] Lie groups and Lie algebras (Gordon and Breach, New York, 1968).

G. Hochschild. [1] The structure of Lie groups (Holden-Day, San Francisco, 1965).

Index